1850–1875

Heyday of the natural-ice trade.

1859

Ferdinand Carré produces a commercial artificial-ice-making machine.

1877

Louis-Paul Cailletet liquefies oxygen and nitrogen, setting the stage for the descent toward absolute zero.

1892

James Dewar develops the thermos bottle.

1898

Dewar liquefies hydrogen at –250°C (23 K).

1900–1915

Alfred Wolff, Stuart Cramer, and Willis Carrier invent commercial air conditioning.

1905

Walther Nernst articulates the third law of thermodynamics, which shows that absolute zero is unreachable.

1908

Heike Kamerlingh Onnes liquefies helium at below 5 K.

1911

Onnes discovers superconductivity in mercury at 4.19 K.

1923

Clarence Birdseye begins mass marketing of frozen foods.

1925

Albert Einstein and Satyenda Bose predict a new state of matter at ultra-low temperatures.

1957

John Bardeen, Leon Cooper, and Robert Schrieffer explain superconductivity.

1980s

Coolants become essential in the production and use of micro-chips and other electronics.

1986

Alex Müller and Georg Bednorz produce "high-temperature" superconductivity.

1995

A Bose-Einstein condensate is created at 170-billionths of a degree above absolute zero.

Absolute Zero
and the
Conquest of Cold

Books by Tom Shachtman

Nonfiction

Absolute Zero and the Conquest of Cold
Around the Block
The Most Beautiful Villages of New England
The Inarticulate Society
Skyscraper Dreams
Decade of Shocks, 1963–1974
The Phony War, 1939–1940
Edith and Woodrow
The Day America Crashed

Collaborative Books

I Have Lived in the Monster (with Robert K. Ressler)
Justice Is Served (with Robert K. Ressler)
Whoever Fights Monsters (with Robert K. Ressler)
Image by Design (with Clive Chajet)
Straight to the Top (with Paul G. Stern)
The Gilded Leaf (with Patrick Reynolds)
The FBI-KGB War (with Robert J. Lamphere)

For Children

The President Builds a House
America's Birthday
Video Power (with Harriet Shelare)
The Birdman of St. Petersburg
Parade!
Growing Up Masai

Novels

Driftwhistler
Wavebender
Beachmaster

Absolute Zero
and the
Conquest of Cold

Tom Shachtman

HOUGHTON MIFFLIN COMPANY

BOSTON · NEW YORK

1999

For information about permission to reproduce
selections from this book, write to
Permissions, Houghton Mifflin Company,
215 Park Avenue South, New York,
New York 10003.

Library of Congress Cataloging-in-Publication Data
Shachtman, Tom, date.
Absolute zero and the conquest of cold / Tom
Shachtman.
p. cm.
Includes index.
ISBN 0-395-93888-0
1. Low temperature research. I. Title.
QC278.s48 1999
536'.56 — dc21 99-33305 CIP

Printed in the United States of America

Book design by Robert Overholtzer

QUM 10 9 8 7 6 5 4 3 2 1

For Mel Berger

Contents

Absolute Zero
and the
Conquest of Cold

Winter in Summer

KING JAMES I OF ENGLAND AND SCOTLAND chose a very warm day in the summer of 1620 for Cornelis Drebbel's newest demonstration and decreed that it be held in the Great Hall of Westminster Abbey. Drebbel had promised to delight the king by making the atmosphere of some building cold enough in summer to mimic the dead of winter, and by choosing the Great Hall the king gave him an enormous challenge, the largest interior space in the British Isles, 332 feet from one end to the other and 102 feet from the floor to the golden bosses of its vaulted white ceiling.

In 1620 most people considered the likelihood of reversing the seasons inside a building impossible, and many deemed it sacrilege, an attempt to contravene the natural order, to twist the configuration of the world established by God. Early-seventeenth-century Britons and Europeans construed cold only as a facet of nature in winter. Some believed cold had an origin point, far to the north; the most fanciful maps represented Thule, a near-mythical island thought to exist six days' sailing north of the northern end of Britain and supposedly visited only once, by Pytheas in the fourth century B.C. — an unexplored, unknown country of permanent cold.

Not until the end of the nineteenth century would a true locus of the cold become a more real destination, as Victorian scientists tried to reach absolute zero, a point they sometimes called "Ultima Thule." Likening themselves to contemporary explorers of the uncharted Arctic and Antarctic regions, these laboratory scientists sought a goal so intense, so horrific, yet so marvelous in its ability to transform all matter that in comparison ice was warm.

In the early seventeenth century, even ordinary winter cold was forbidding enough that the imagination failed when trying to grapple with it. "Natural philosophers" could conceive technological feats that would not be accomplished until hundreds of years later — heavier-than-air flight, ultrarapid ground transportation, the prolongation of life through better medicines, even the construction of skyscrapers and the use of robots — but not a single human being envisioned a society able to utilize intense cold to advantage. Perhaps this was because while the sources of heat were obvious — the sun, the crackle of a fire, the life force of animals and human beings — cold was a mystery without an obvious source, a chill associated with death, inexplicable, too fearsome to investigate.

Abhorrence of cold was reflected in only sporadic use made of natural refrigeration, an omission that permitted a large percentage of harvested grains, meats, dairy products, vegetables, fruits, and fish to spoil or rot before humans could eat them. And since natural refrigeration was so underutilized, producing refrigeration by artificial means was considered a preposterous idea. No fabulist in 1620 could conceive that there could ever be a connection between artificial cold and improving the effectiveness of medicine, transportation, or communications, or that mastery of the cold might one day extend the range of humanity over the surface of the earth, the sky, and the sea and increase the comfort and efficiency of human lives.

How did water become snow in the heavens or ice on the earth? What formed the snowflakes? Why was ice so slippery? In 1620 these and dozens of other age-old, obvious questions about the cold were considered not only unanswerable but beyond the reach of investigation. Cold could neither be measured, nor described as other than the absence of heat, nor created when it was not already present — except, perhaps, by a magician.

On that summer day when the king and his party approached Westminster Abbey — which was in need of some repair, the fabrics torn, the buttresses on the northwest side crumbling in places — James Stuart was getting on in years, having recently passed his fifty-fourth birthday. In middle age he was still short, broad-shouldered, and barrel-chested, but his hair, once dark, had thinned to a light brown, and the rickets that had affected his growth in youth had lately made his gait more uneven and erratic, requiring him as he walked to lean on a companion's shoulder or arm. He suffered from sudden attacks of abdominal pain, rheumatism, spasms in his limbs, and melancholy. After the loss of his queen, Anne of Denmark, in 1619, he had begun to do uncharacteristic things: even though the king and queen had been estranged and had lived separately for years, James honored Anne in death by siting her sepulcher in Westminster, near the last resting place of his mother, Mary, Queen of Scots. Very few sepulchers or honorary statues decorated the abbey just then.

Summer played havoc with the king's delicate skin, described as "soft as taffeta sarsnet," thin, fragile, and subject to frequent outbreaks of itching and to sweating, which exacerbated the itches. He also suffered from a sensitivity to sunlight so severe that undue exposure to the sun overheated him to the point of danger. His susceptibility to heat was worsened by the thick clothing he habitually wore and the doublets specially quilted to resist knife thrusts, an augmentation deemed necessary after several assassination attempts against him. "Look not to find the softness of a down pillow in a crown," the king had written earlier that year, in a small book of meditations on the biblical verse about Jesus crowned with thorns, "but remember that it is a thorny piece of stuff and full of continual cares."

Aside from obtaining relief from the heat, James's interest in the coming demonstration derived from his lifelong obsession with witchcraft and unnatural matters, given fullest flower in his book *Demonologie*, published in 1597. In 1605, two years after James had

ascended to the throne of England upon the death of Queen Elizabeth, his fascination with the occult and his continual search for
entertainment led him to accede to an entreaty for patronage by the
Dutchman Cornelis Drebbel. James installed Drebbel and his family,
with room and board and a grant for expenses, in a suite at Eltham
Palace so that Drebbel could set up a laboratory and manufacture,
for the particular delight of James's son Henry, such devices as a
"perpetual-motion" apparatus, a self-regulating oven, a magic lantern, and a thunder-and-lightning machine.

That Drebbel billed himself to James as a magician, not a scientist,
shines through in a letter the Dutchman sent home in 1608, regarding his magic-lantern display:

> I take my stand in a room and obviously no one is with me. First I
> change the appearance of my clothing. . . . I am clad first in black
> velvet, and in a second, as fast as a man can think, I am clad in green
> velvet, in red velvet, changing myself into all the colors of the world
> . . . and I present myself as a king, adorned in diamonds, and all sorts
> of precious stones, and then in a moment become a beggar, all my
> clothes in rags.

In front of his audience, Drebbel appeared to change into a lion, a
bird, a tree with trembling leaves; he summoned ghosts, first the
menacing kind, then heroic spirits such as Richard the Lionhearted.
Given Drebbel's apparent ability to produce thunder and lightning
at will, and to change shapes, it was no wonder that some in his
audience deemed him godlike.

The precise date of Drebbel's 1620 demonstration of the power of
cold, the identities of those present at it, and the efficacy of the
cooling went unreported by eyewitnesses. We have only secondhand
accounts of it. But reasoned guesses based on other known information may shed additional light on the event. It probably occurred
after July 12, the installation date of John Williams as dean of the
abbey, replacing a long-serving, more conservative dean. Williams
of Salisbury was a progressive of sorts and more likely than his

predecessor to have acquiesced to Drebbel's display in the hallowed abbey. Moreover, he had been chosen as dean by George Villiers, then the marquis of Buckingham, King James's last and most influential homosexual lover. Buckingham was likely to have been in the small crowd that day; he and the king shared a fondness for magic, alchemy, and surprising mechanical apparatuses. To arrange his own entertainments, Buckingham employed on his estate a young man from Antwerp named Gerbier, who in all probability was likewise in attendance at Westminster, perhaps as an assistant to Drebbel; two years earlier, Gerbier had praised Drebbel in an elegy on the death of Drebbel's brother-in-law, which suggests a working relationship between the Dutch expatriates. Other guests may have been the astrologer and crystal gazer John Lambe, whose influence at court was considerable, and Salomon de Caus, maker of fantastic fountains and spectacular gardens, who had earlier worked alongside Drebbel in the royal service. Assisting Drebbel were, in all likelihood, Abraham and Jacob Kuffler, Dutch brothers who had come to England that year, begun apprenticeships with him, and concocted a scheme in which one or the other would marry Drebbel's daughter and thereby become privy to his marvelous secrets.

So: Probably in the afternoon, when the heat of the day was at its height, and between one of the seven daily sessions of monks' devotions, the royal party entered the abbey, presumably through a side door to the north portal — opening the great north-portal doors would have spoiled everything — and stood in the shadowed edifice, to be welcomed by one of the most mysterious men of his time. Many in England believed, with Ben Jonson, that Drebbel was a mountebank, a charlatan, and possibly a necromancer. Some in Holland called Drebbel *pochans* or *grote ezel,* "braggart" or "big jackass," but there were as many others, in both countries, who respected Drebbel as an inventive genius because he had astonished them with some marvelous devices.

Born at Alkmaar in the north of Holland in 1572 to a landowning family, Cornelis Jacobszoon Drebbel had little formal schooling. For many years he remained unable to read or write in Latin or English, and even after he had taught himself both languages, he continued to despise books and wrote little. In his teens he apprenticed in nearby Haarlem to Hendrik Goltzius, an engraver who dabbled in alchemy, and later married Goltzius's sister. He also evidently learned some technical matters from two Haarlem brothers who later became well known for innovations in mathematics and optics. In 1598 Drebbel was awarded patents for a water-supply system and for a form of self-winding and self-regulating clockworks. In 1604 he published *On the Nature of the Elements,* a short treatise confabulating alchemy, pious thoughts, and speculation about the interpenetration of the four elements — earth, fire, air, and water. In 1605 Drebbel wrote to James of England, promising him the greatest invention ever seen, a *perpetuum mobile,* a perpetual-motion machine, and dedicating to the king the English edition of his book on the elements.

The device Drebbel made at Eltham did not produce perpetual motion, of course, since that is impossible, but according to the contemporary account of Thomas Tymme, a professor of divinity who thought it wondrous, this was a clock with a globe, girdled with a crystal belt in which water was contained, accompanied by various indicators that told the day, month, year, zodiac sign of the month, phases of the moon, and rise and fall of the tides. In Tymme's eyes, Drebbel's machine reflected the perpetual movement of the universe, set in motion by the Creator. Tymme reported in a book that when King James had seemed unwilling to believe in its perpetual motion, Drebbel, that "cunning *Bezaleel,* in secret manner disclosed to his maiestie the secret, whereupon he applauded the rare invention." Though Tymme said the machine was operated by "a fierie spirit, out of the mineral matter," most likely it was powered either by variations in atmospheric air pressure or by the expansion and contraction of heated and cooled air.

By 1610 the fame of "the philosopher of Alkmaar" had reached the court of Rudolf II, emperor of Bohemia, who invited Drebbel and his family to Prague, where Drebbel would have opportunity to replace the former wizard of the castle, the noted English alchemist Dr. John Dee. Rudolf had earlier lured Danish astronomer Tycho Brahe to the castle at Hradschin, but by this era the emperor had gone beyond such true scientists and was neglecting the affairs of state to work alongside his invited artificers in an effort to find the elusive philosophers' stone, a substance that alchemists believed would transmute base metal into gold. Drebbel's adventure in Prague ended in disaster: Rudolf died in 1612 and his successor imprisoned the Dutchman, either for his loyalty to the wrong faction or for his alleged involvement in a scheme to embezzle money and jewels. Drebbel wrote an impassioned letter to King James in 1613, promising not only a new and improved self-regulating clockwork but also "an instrument by which letters can be read at a distance of an English mile" as well as an elaborate fountain featuring curtains and doors that opened at the touch of the sun, water flowing on cue, and music playing automatically on small frameless keyboards, while "Neptune would appear from a grotto of rocks accompanied by Tritons and sea-goddesses." The king forthwith sent Drebbel instructions to return to England and money for the journey.

Drebbel made that fountain for King James, along with a camera obscura and a crude telescope. As time went on, pressure grew on him to continue to produce magically ingenious if not miraculous devices in exchange for his supper, especially after 1618, when circumstances combined to spur James to submit to a new regime of austerity and curb his prodigious household spending.

In 1620 Cornelis Drebbel was forty-eight, and although his beard had turned gray he was still the "fair and handsome man . . . of gentle manners" that a visiting courtier had described years earlier; the Dutch poet and scientist Constantijn Huygens, a recent acquaintance, thought he looked like a "Dutch farmer" but one full of "learned talk . . . reminiscent of the sages of Samos and Sicily." Drebbel's genteel reputation was often contrasted with that of his wife, Sophia, who according to another account spent all of Drebbel's income "on the entertainment of sundry lovers." Huygens's parents warned him against associating with this "magician" and "sorcerer" — but still asked their son to find out about lens-grinding techniques from him.

At the time of the cold demonstration, according to Drebbel's assistants, the inventor lived "like a philosopher," oblivious to fashion, despising the world and especially its great men, caring for naught but his work, willing to talk only to those who shared his fondness for tobacco, often neglecting to eat because he was lost in scientific thought. These were the circumstances that led him to devise a triumph of man over nature, the reversal of the seasons, the creation of winter in summer.

When the king and his followers entered the abbey that summer day, probably through a door beneath the great rose stained-glass window, they were likely ushered to a section near the center, the sacrarium, a relatively narrow and shorter enclosure within the larger hall. There the air was, as Drebbel had promised, quite cool. All would have felt the chill to one degree or another. Guests would have looked askance at certain troughs and other devices they could not fathom, placed near the bases of the walls, and perhaps for guidance up to the white ceiling, partially blackened with soot from the tens of thousands of candles burned in the chamber over the centuries. Shortly, because of James's overheated condition and near-continual sweating, the king began to shiver and he retreated outside, followed by the rest of his party. The demonstration was a success.

How did Drebbel do it? Since he left no written description, and the few accounts of the event are secondhand, answering the question requires some lateral analyses. Years before the incident at Westminster Abbey, the engineer and dramatist Giambattista della Porta had produced ice fantasy gardens, intricate ice sculptures, and iced drinks for Medici banquets in Florence; the excited reports by the nobility about these feats spread through Europe and can be found today in letters and memoirs. Of the more reliable reporters of Drebbels's feat, only Francis Bacon made reference in a 1620 book to "the late experiment of artificiall freezing" at Westminster, so there is a decided lack of detail about the demonstration of mechanical air conditioning, though it was stark evidence that people could exert mastery over a condition of nature.

The lack of notice was consistent with a general failure to take Drebbel's remarkable demonstration seriously. To contemporaries, this must have seemed just another piece of magic at a time when the elite of society were struggling to free themselves from a fascination with the more-than-natural that had held the world in thrall for a thousand years. Magic and "natural science" then coexisted uneasily, and it was far from certain that science would eventually prevail. Drebbel's "experiment" may also have failed to attract more attention because of its lack of immediate practical application.

Considerably more astonishment was professed at Drebbel's well-reported 1621 demonstration of a submarine. In three hours the boat traveled "two Dutch miles" underwater on the Thames, from Westminster to Greenwich, in front of the king and thousands of onlookers. None could figure out how the submerged crew of twelve — plus the inventor himself, who risked drowning along with them — could continue to breathe in the absence of fresh air. Drebbel provided a clue to the submarine's air supply in his *Fifth Element*, published that year, which included the cryptic statement that "saltpetre, broken up by the power of fire, was thus changed into something of the nature of the air." Scientific analysis was so rare in 1621 that no one picked up on that clue; decades later British chemist and physicist Robert Boyle would partially comprehend what this demonstration accomplished, writing that "Drebbel conceived that it is not the whole body of the air, but a certain quintessence . . . or spirituous part of it that makes it fit for respiration," and figuring out that when Drebbel observed that the air in the submarine was becoming exhausted, "he would by unstopping a vessel full of his liquor speedily restore [to] the troubled air such a proportion of the vital parts, as would make it again, for a good while, fit for respiration." In short, Drebbel had isolated and discovered oxygen, 150 years before Joseph Priestley. But today Drebbel's name is nowhere associated with that major advance in chemistry.

Drebbel's fondness for the dramatic presentations of the magician rather than the steady progress of the scientist may also help explain, in part, why his preternatural stunt of cooling Westminster in summer produced few reverberations. An inventor and court entertainer, he felt keenly the need to keep the secrets of his demonstrations to himself, a need reflected by his lifelong refusal to document and publish his experiments properly or to keep a diary. "Had Drebbel compiled notebooks describing his undoubted technological works," writes L. E. Harris, president of a society dedicated to the history of engineering, "he might have attained some lasting fame even without having an influence on future technologies, as is the case with Leonardo da Vinci." In the time-honored way of the magician, Drebbel vouchsafed his "secrets" only in fragments to his apprentices, the voracious Kufflers — but evidently he did not tell them very much, for after Drebbel's death they were not able to replicate his feats, though they made money from a dye works based on his "secret" formula.

Drebbel appears to have been convinced that if he disclosed the secrets of his work, he would lose the aura of mystery that made him attractive to the king; moreover, by retaining the secrets, he affected to possess a power over nature that in some measure counterbalanced the power of the king over ordinary mortals. But this was only posturing. How dependent Drebbel was became obvious only when King James's death removed his stipend, which reduced him to what Flemish artist Peter Paul Rubens wrote was an "extraordinary" appearance of such shabbiness and disarray that it "fills one with surprise."

Drebbel's refusal to reveal his secrets was accepted and sealed by his audience's equal reluctance to demand explanations for marvelous devices and demonstrations. Heinrich van Etten, a contemporary, suggested that audiences found mathematical and scientific puzzles more entertaining if their inner workings were concealed, "for that which doth ravish the spirits is an admirable effect whose cause is unknowne, which if it were discovered, halfe the pleasure is lost." The statement reflects a lack of curiosity that ran throughout society at that time, from the basest peasant to the highest noble.

Today we believe curiosity is central to science and perhaps to all of human progress; curiosity is the engine that drives the intellect to seek the causes of things. "Curiosity is one of the permanent and certain characteristics of a vigorous mind," Samuel Johnson would write in 1751, and few could disagree with him.

But in 1620 prevailing opinion disparaged curiosity. The distaste rested on two pillars of ancient thought that resonated throughout the late medieval and Renaissance eras. In the fifth century Saint Augustine had condemned curiosity as a base longing to know the trivial, contrasting it with the elevated pleasures of faith, which he believed provided all the explanations that humankind needed; curiosity was anathema because it meant delving too deeply into what God had created. Adding to the distrust of curiosity and of any quest to unlock the "secrets" of natural phenomena was a belief that investigating nature's hidden workings ran counter to Aristotle's teachings, inscribed nearly a thousand years before Augustine. Aristotle had taught that nature could be entirely apprehended by the senses, that knowledge was not obtainable through experiment and could be derived *only* as a byproduct of reason and logic. In the thirteenth century, Thomas Aquinas had fused the philosophies of Aristotle and Augustine, as they related to scientific inquiry, and since then his synthesis had been dominant. John Donne, who owed his high ecclesiastical position to King James, vehemently agreed with Aquinas that it was impious to attempt to uncover any hidden truths about nature.

In the early 1600s, however, beliefs that decried curiosity and restricted information about the "secrets" of nature to a handful of cognoscenti were under attack, and the most highly influential Eng-

lish opponent of such views was a man who tried to explain Drebbel's demonstration at Westminster, though he probably had not
been present at it: Sir Francis Bacon, Baron Verulam, lord chancellor
of England. Lawyer, historian, philosopher, and politician, Bacon
more than anyone else in England helped banish magic and secrets
by championing science based on experimentation. Constantijn
Huygens might write of Drebbel and Bacon in the same sentence
and contend that their accomplishments were of equal moment, but
they were not colleagues. Rather, they were polar opposites, Drebbel
among the last of the magician-artificers and Bacon the first true
English scientific thinker. In Drebbel's refusal to explain his stunt
and Bacon's insistence on trying to discern its chemical mechanism
of cooling lies the deeper significance of Drebbel's demonstration: it
symbolized the passing of the era in which magic held all the fascination and the arrival of science at center stage to begin the process
of providing explanations of nature that would greatly advance human civilization.

We infer Bacon's absence at the Westminster event because he did
not write himself an immediate note about it, as he had done after
viewing Drebbel's demonstrations of earlier devices and machines at
Eltham Palace. Bacon's appetite for scientific stunts was declining;
in 1605, while courting King James, he had condoned the study of
marvels, witchcraft, and sorcery "for inquisition of truth, as your
majesty has shown in his own example [in *Demonologie*]," but later
Bacon insisted that "experiments of natural magic should be sifted
diligently and severely before they are received, especially those . . .
commonly derived . . . with great sloth and facility both of believing
and inventing."

Another likely reason for Bacon's absence was the gathering
storm, fomented by his political enemies, that within a year would
result in his abject fall from favor. Shortly after James made Bacon
viscount of St. Albans in early 1621, the nobleman was impeached for
accepting bribes; after confessing to his guilt, he was stripped of his
position and banished from London, though he was spared incar-

ceration. The deeper reason for Bacon's eclipse was related to his growing advocacy of experimental science. English scientist Robert Hooke later identified that reason, in comparing Bacon's treatment to that of Italian scientist Galileo by the Inquisition: "Thus it happened also to . . . Lord Chancellor Bacon, for being too prying into the then receiv'd philosophy."

Bacon was never a man to ignore what another experimenter might turn up that could be relevant to his own studies, and perhaps that is why, in *Novum Organum,* published later in 1620, he wrote the short section that, according to an associate, tried to fathom "the late experiment of artificiall freezing" at Westminster: "Nitre (or rather its spirit) is very cold, and hence nitre or salt when added to snow or ice intensifies the cold of the latter, the nitre by adding to its own cold, but the salt by supplying activity to the cold of the snow." Nitre, also known as saltpeter, is a common chemical compound (today called potassium nitrate) and the active ingredient of gunpowder. Bacon's guess about Drebbel using nitre was a good one: the court artificer had himself written of saltpeter and was also on intimate terms with Sir Thomas Chaloner, author of a book solely about nitre; moreover, as Bacon hints, many alchemists and would-be scientists had been experimenting with the cold-inducing aspects of nitre and common salt.

A source for those experiments was one of the most popular "books of secrets" of the age, Giambattista della Porta's *Natural Magic,* first published in Italy in 1558 and enlarged — as well as translated into virtually every other European language — in 1589. Della Porta was one of the most famous men in Italy, a friend of German astronomer Johannes Kepler and Galileo, a man so learned in the ways of nature that he was expected at any moment to discover the philosophers' stone. Jailed by the Inquisition for his magic, he continued to write about it. In *Natural Magic,* following sections treating alchemy, invisible writing, the making of cosmetics, gardening, and the accumulation of household goods, della Porta appended a final miscellany, "The Chaos," in which he mentioned mixing snow and nitre to produce a "mighty cold" that was twice as cold as either substance — cold enough to make ice.

With these hints, and some technology of the era, we can finally reconstruct how Drebbel probably accomplished his feat.

At an early hour of the morning, Drebbel and his assistants brought into Westminster Abbey long, watertight troughs and broad, low vats and placed them alongside the walls and in the midst of the limited part of the abbey that they planned to cool, most likely that inner, narrow transept near the portal through which the king and courtiers would enter, an area they knew would be in shade most of the day and especially at that hour. They also brought in snow, which would have been available from those among the nobility who had on their estates underground snow pits to keep unmelted snow and ice in storage after the winter, to use for cooling drinks in summer. Drebbel filled the troughs and vats partway with water, the coolest he could find, which he no doubt had fetched directly from the nearby Thames. For several hours, he infused nitre, salt, and snow into the water, creating ice crystals and a mixture whose temperature — if he could have measured the temperature, which he could not, since no thermometers capable of such accuracy yet existed — was actually reduced *below* the freezing point of water, as della Porta had guessed. Some of the troughs were metal, and the freezing mixture chilled the metal, which aided the refrigerating process by keeping the contents of the troughs cold.

More to the point of the exercise, the freezing mixture cooled the air directly above the troughs and vats. In Drebbel's *Elements* treatise he referred to the frequently observed phenomenon of heated air rising, and he seems also to have understood that cool air is heavier than warm air and tends to stay close to the ground. Now he used this principle to generate a mass of cool air that displaced warmer air in the cathedral up in the direction of the capacious ceiling. He did not need to force the warm air to rise very far — just 10 feet high or so, until it was above the height of the king and courtiers. And he did not need to make the space very cold — a decrease in temperature from, say, 85° to 65°F would have proved sufficient to chill an overheated king. This cooling Drebbel accomplished over the course of several hours, perhaps aiding the process by fanning the cool air so that remaining pockets of warm air thoroughly dispersed, before the court party arrived and experienced the shock of the cold.

Exploring the Frontiers

I N THE SEVENTEENTH CENTURY, the capital cities of Europe
and England were enlarging in size and population relatively
slowly, in part because of society's limited ability to provide food
to locations that could not grow enough to feed their own residents.
A quarter of the grains, fruits, and vegetables would rot in the fields
before being harvested, and eggs and milk would quickly spoil. If
the destination of the crops or dairy products was more than a
day's wagon ride from the farm, another fraction might become
inedible during transport. Evidence that farmers knew that cold
retards spoilage comes from their general practice of bringing food-
stuffs to the city at night, to take advantage of lower evening temperatures. At city markets, animals used for food were generally
killed only after customers had bought them, or no more than a few
hours before sale, because uncooked or untreated flesh would not
remain edible for long. To hold the live animals, butchers required
larger premises than other shopkeepers, which raised the cost of
their meat.

Owing in large measure to the absence of refrigeration, fresh
meats, fish, milk, fruits, and vegetables made up a lower percentage
of the diet than bread, pickled vegetables, cheeses, and preserved
meats. A great deal of ingenuity went into preserving by pickling in
salt or sugar, smoking, drying, or excluding air by submerging foods
in oil, all of which substantially altered the character and taste of

produce or meat. Vegetables and fruits could not be obtained out of season, except at inordinate cost or under special circumstances, as when a king would dispatch a ship to Morocco to bring back oranges in winter.

In the Temperate Zone, even when ice was available, it was not extensively used for food preservation, the nobility employing their ice facilities mainly to provide chips to cool their wine in summer, much as the ancient Romans had done. Seventeenth-century technology for utilizing cold had not advanced one whit over that of ancient times. Pliny ascribed to Emperor Nero the invention of the ice bucket to chill wines, designed to eliminate the need to drink wine diluted by ice that had been stored in straw and cloth. Zimrilim, ruler of the Mari kingdom in northwest Iraq around 1700 B.C., built a *bit shuripin,* or icehouse, near his capital on the banks of the Euphrates. In China, the maintenance of icehouses for the preservation of fruits and vegetables dates to the seventh century B.C.; a book about food written during the Tang dynasty (A.D. 618–907) referred to practices begun during the Eastern Chou dynasty (770–256 B.C.), when an "ice-service" staff of ninety-four people performed the tasks of chilling everything from wine to corpses. In the fourth century A.D., the brother of the Japanese emperor Nintoku offered him ice from a mountain, a gift so charming that the emperor soon designated the first of June as the Day of Ice, on which civil and military officials were invited to his palace and were offered chips, in a ceremony called the Imperial Gift of Ice.

Night cooling by evaporation of water and heat radiation had been perfected by the peoples of Egypt and India, and several ancient cultures had partially investigated the ability of salts to lower the freezing temperature of water. Both the ancient Greeks and Romans had figured out that previously boiled water will cool more rapidly than unboiled water, but they did not know why; boiling rids the water of carbon dioxide and other gases that otherwise retard the lowering of water temperature, an explanation the Greeks and Romans were unable to reach or understand.

Progress in the use of cold had been held back by a dearth of basic knowledge about its physics and chemistry. The advance of such knowledge in turn depended on social change, which after a thousand years of stasis had become the order of the day in the seventeenth century. Partly owing to the Protestant challenge to Catholicism, partly to the discovery of the Americas, many thinkers embraced the radical notion that there was more to the world, and to knowledge, than had previously been believed.

This was no minor shift in emphasis but a sea change in society, writes historian of ideas Barbara Shapiro, in which the practitioners of law, religion, and science all became "more sensitive to issues relating to evidence and proof. . . . Experience, conjecture, and opinion, which once had little or no role in philosophy or physics, and probability, belief, and credibility . . . now became relevant and even crucial categories for natural scientists and philosophers." Christiaan Huygens, the mathematically gifted son of Constantijn and the inventor of the pendulum clock, expressed the new understanding: "'Tis a Glory to arrive at Probability. . . . But there are many degrees of Probable, some nearer Truth than others, in the determining of which lies the chief exercise of our judgment."

Cornelis Drebbel cared little for the glory of probability; he wanted to make a living. After demonstrating his power over the cold at Westminster, he made no further public displays of low temperatures, perhaps because he garnered no encouragement for them, in the form of either honor or money. The submarine demonstration did bring him employment, though, and after the death of King James in 1625, Drebbel worked with the military, helping to manufacture explosives, which he took into battle in several Buckingham-led naval expeditions against France. During these forays, he was to be paid at the high rate of £150 a month to set fire to the enemy. The expeditions failed, and Drebbel was unable to collect his pay for the last one. He tried in vain to revive a scheme to distribute heat to the houses of London via underground pipes, and he was part of an unsuccessful attempt to drain fens to make arable land. Desperate for income, he started a brewery and alehouse near London Bridge, attracting attention with an underwater contraption that appeared to be a monster. Drebbel died in 1633, and the secrets of his marvelous devices perished with him.

For Francis Bacon, the glory of arriving at probability became the touchstone of his later life. Shorn of his political responsibilities, he turned his mind again toward natural science, writing several seminal works during his last five years of life, from 1621 to 1626. It was in these natural science books, perhaps more than in his earlier political tracts, that Bacon did what Robert Hooke later admired: he countered "the receiv'd philosophy" and in so doing made possible many subsequent steps in science, in particular those leading to the greater understanding of the cold that this book chronicles.

The most formidable barrier to comprehending cold was established belief, and Bacon's intellectual leadership was crucial to piercing this barrier. His lifelong aim was to be "like a bell-ringer, which is first up to call others to church." Whether exampled by the parish of law or the parish of natural philosophy, for Bacon the goal was "the study of Truth," pursued through the "desire to seek, patience to doubt, fondness to meditate, slowness to assert, readiness to reconsider, carefulness to dispose and set in order." He applied these virtues in the service of the inductive method, the making of proper observations and experiments as a basis for drawing conclusions about the workings of the natural world. His *Instauratio Magna* announced a "trial" of the "commerce" or correspondence between what humankind believed it knew about the natural world and the true "nature of things," because the goal of bringing the two into congruence was "more precious than anything on earth." To properly contemplate the natural world, he contended, required the rejection of error-riddled previous natural philosophies, particularly that of Aristotle, whose natural philosophy Bacon thought overly based on deductive logic. "I seem to have my conversation among the ancients more than among those with whom I live," Bacon explained in a letter to a friend in Paris, the chemist Isaac Casaubon.

In Aristotle's view, if one knew the significant "facts" about nature — such as that all things were combinations of the four elements, air, fire, earth, and water — one could deduce whatever humanity needed to know about the world. Aristotle's seventeenth-century followers refused to consider as valid the contemporary experiments investigating or manipulating nature to determine previously hidden properties and causes. Bacon supported such experiments, arguing that "nature exhibits herself more clearly under the trials and vexations of art [forced experimentation] than when left to herself," since nature was like Proteus, the mythical creature who could conceal his identity in myriad shapes until bound in chains, whereupon his true identity was revealed. While Bacon's main target was Aristotle, he also sought to refute artificers such as Drebbel, whose dabblings were based on inconsistent observations and on an absence of rigorous, documented experimentation. "My great desire is to draw the sciences out of their hiding-places into the light," Bacon also told Casaubon. The public considered things to be "marvelous" only so long as their causes remained unknown, he wrote, but "an explanation of the causes removes the marvel," and the business of science must be to identify and explain those causes.

For the mind to pursue a better understanding of nature, Bacon believed that it must first be purged of preconceptions. Identifying four "idols" of preconception, he railed against them as though he were Jehovah warning his chosen people against the worship of false gods. These were the Idols of the Theatre, a reliance on received philosophical systems, which had perverted the rule of demonstration — that was Aristotle's failing; the Idols of the Tribe, which distorted truth by stressing the correctness of one's own tribe's ideas over those of others; the Idols of the Cave, which prevented individuals from seeing their own defects (principally produced by poor education), so that they looked for sciences "in their own lesser worlds, and not in the greater or common world"; and the Idols of the Marketplace, which used words to deceive the mind, to trick it into thinking that night was day. All these stood in the way of proper research on the cold.

As antidote to the Idols, a year before his death, Bacon put aside other writings to inscribe, almost in one sitting, a fable of the scientific ivory tower of the future. *The New Atlantis* was Bensalem, a city on a tropical island that was an unmistakable contrast to Augustine's faith-based "city on a hill." The "lanthorn" (lantern) of this civilization was Salomon's House, run by an "Order . . . dedicated to the study of the Works and Creatures of God," an institution alternatively known as the College of the Six Days' Work. The college was organized along the lines of houses of higher learning that Bacon had wished to establish in England, but its laboratories and the attempts of the Bensalemites to command nature bore a distinct resemblance to the facilities and constructions of Cornelis Drebbel and to those of Salomon de Caus, who had designed fantastic gardens for King James.

In the tale, mariners sailing from Peru became lost in a storm and sought shelter and medical assistance on the island; there the group learned about the work of Salomon's House from one of its elders, a majestic figure whose gaze "pities men." There were vaults, furnaces, laboratory workhouses, and 3-mile-deep caves used for "all coagulations, induracions, refrigerations, and conservations of all natural bodies." Half-mile-tall towers with telescopes allowed observations of "diverse meteors . . . winds, rain, snow, hail" and had "engines" for multiplying these natural forces. There were gardens for grafting, and mechanical shops and engineering facilities to build faster means of locomotion and better instruments of war, and to scientifically investigate the motion of birds, so flying machines could be made. The experimenters investigated and imitated all natural phenomena — and then, having understood how nature works, they made flowers bloom out of season and forced water to become ice. The aim of such experiments was to gather data for theorists who would draw "axioms" from it and construct a coherent natural philosophy. The relative weight Bacon gave research and theorizing was displayed by the division of labor at Salomon's House: thirty-three experimenters performed their duties, and just three elders of the community analyzed the experimental results. Beyond distilling the "knowledge of Causes, and the secret motion of things," Salomon's House aimed at "the enlarging of the bounds of Human Empire, to the effecting of all things possible."

After King James's death in 1625, Bacon was permitted to reside occasionally at Gray's Inn in London, rather than having to remain a dozen miles from the city; and there were further indications that the change in monarchs might completely end his banishment and once again bring him to counsel the crown. In March 1626, while riding in a coach with the physician to the new king, Charles I, Bacon looked at the snow covering the ground and decided to try an experiment to see whether it would preserve the flesh of an animal as well as salt did. That he would even consider such a test is added evidence that at this time natural refrigeration was not generally used for animal flesh. To conduct an experiment was an unusual act for Bacon, whose books mainly featured his analysis of others' work; perhaps writing *The New Atlantis* spurred him to take a more active role in the investigative process. In any event, he and the physician stopped the carriage near Highgate to go into a poor woman's house and buy a chicken from her, which she quickly dispatched and cleaned at their request. Then the two men returned outside, bent down to the ground, gathered snow, and stuffed and wrapped the carcass with it.

The snow so chilled Bacon, his onetime secretary Thomas Hobbes later recalled to John Aubrey, who recounts his story in his well-known *Brief Lives*, that Bacon became too ill to travel and was rushed to the nearby home of the earl of Arundel — the earl then being absent, imprisoned in the Tower of London. Bacon was put into a bed warmed by a pan, but it was a damp bed that had not been used for a year, and his condition worsened. He wrote Arundel, explaining what had happened and citing the ancient story of Pliny the Elder, the Roman historian whose inquisitive sense had drawn him too close to Vesuvius, where the volcanic eruption killed him. Bacon knew he was dying, but in this letter he commented that his experiment into the ability of snow to preserve the flesh of the chicken "succeeded excellently well." Hours after writing this wry note to his host, on Easter Day 1626, Sir Francis Bacon died of pneumonia.

"In the generation after Bacon's death, many men called themselves Baconians who grasped only the details of his work," writes historian Hugh Trevor-Roper, concluding that "it is the fate of all great men to be quickly vulgarized." The Puritans seized upon Bacon's notion that knowledge must be used for the improvement of human welfare, but they refused to equally honor his other point, that experimentation is required to advance humankind's store of knowledge. During the two civil wars and the dominance of Oliver Cromwell in England, in the 1640s and 1650s, little that challenged orthodox views in any arena was tolerated. Only near the end of the Cromwell era, in the late 1650s, did true Baconianism in science resurface, in the formation of a loose cohort of scientific experimenters pledged to Baconian ideals, some of whom met first at Gresham College in London and later at Oxford.

United in their emphasis on the need to experiment and to push aside the Aristotelian-Augustinian-Aquinan way of apprehending and explaining the world, they took their philosophic cues from Bacon, and from Copernicus as interpreted by Galileo. Aristotle had watched the sun rise in the east and set in the west and deduced the logical conclusion that the sun revolved around the earth; Copernicus and Galileo observed and measured the movements of a greater number of astronomical bodies, applied the tenets of mathematics, and inductively concluded that it was highly probable that the earth revolved on its axis daily and around the sun annually.

The new "invisible college" group agreed with and admired this startling Galilean conclusion about nature and adopted the mostly inductive method by which it had been reached. Robert Boyle, Robert Hooke, Christopher Wren, and others began to evolve ways of separating valid from spurious experimentation, in the process accelerating the dissolution of magic's thousand-year spell over the realm of explanations of natural phenomena. "Mountebanks desire to have their discoveries rather admired than understood," Boyle charged, but "I had much rather deserve the thanks of the ingenious, than enjoy the applause of the ignorant."

To deserve the thanks of well-informed people, a scientist's experiments had to be performed in a public though restricted space, before an audience composed of people whose level of knowledge was high enough to properly assess the scientific method and results, but who would not too quickly assert causal explanations of what they saw. And furthermore, for experimental results to be deemed conclusive, the experimenter first had to write his procedures in precise and understandable terms, so that others could replicate the experiment and its results. These Baconian precepts became the basis for establishing in the 1660s the ideal audiences, witnesses, and venues for scientific experimentation: the Royal Society in England and the Académie Royale des Sciences in France, arenas in which scientific work on cold would be judged.

Though the French institution was modeled on the British one, it was more Baconian, because it was more rigorous in selecting its members, and those members, once elected, were given stipends by the government to devote their energies solely to science. The Royal Society's Fellows had to make their own livings, and had to tax themselves to buy scientific equipment.

Boyle and other leading Fellows of the Royal Society were able in their various studies to accomplish much more, scientifically, than Bacon himself, in large measure because Bacon had smoothed their path by erasing belief as a barrier to discovery. "The works of God are not like the tricks of jugglers or the pageants that entertain princes, where concealment is requisite to wonder; but the knowledge of the works of God proportions our admiration of them," Boyle could contend. Hooke could express a similar rationale in his ecstasy at finding in his microscope's view natural forms "so small, and so curious, and their design'd business so far remov'd beyond the reach of our sight, that the more we magnify the object, the more excellencies and mysteries do appear. And the more we discover the imperfections of our senses, and the Omnipotency and Infinite perfections of the great Creator."

The seventeenth-century experimenter who did the most extensive work in the arena of the cold was Robert Boyle. Born a year after Bacon's death, Boyle was the youngest son of an extremely wealthy man, the earl of Cork. In an autobiographical note, Boyle guessed he was the thirteenth or fourteenth child of a mother who died of consumption soon after she gave birth to him, and of a father who died by the time "Robin" was seventeen. He never attended a university, but he studied with private tutors at home and on the Continent. Sickly as a youth, badly injured by what he described as a fall "from an Unruly horse into a deep Place," he was troubled in his adulthood by kidney stones, weak eyesight, and "paralytic distemper."

Though he was one of the first experimental scientists, Boyle was by modern standards still in thrall to magic and irrational disciplines. He lobbied for repeal of an old law against alchemy because such prohibitions restricted legitimate chemical research, but he was not beyond experimenting with and extolling the healing properties of human and horse manure, was convinced of the medicinal value of ground-up millipedes, and believed astrologically based notions such as that grape juice stains could be washed out of garments most readily at the season when grapes ripened on the vine.

Boyle initially began his research in the areas of agriculture and medicine but then gravitated to physics and chemistry. Oxford colleagues ribbed him for pursuing chemistry, which "professed to cure no disease but that of ignorance." His chemical research also incensed the Dutch rationalist philosopher Benedict Spinoza, who castigated Boyle in letters to the Royal Society for subordinating reason to experiment and for believing that a chemical combination of particles could act differently than a physical mixture.

Boyle's large private fortune enabled him to spend liberally to support his studies. No other natural philosopher in England, it was said, could afford the expense of constructing and testing the first — and for some time, the only — vacuum apparatus in existence on the island, fabricated for him by Hooke. Boyle's work on "the spring of the air," published in 1660, secured his scientific reputation. Vacuums led him to conduct research on air pressure, from which he deduced Boyle's law, that the volume of a given amount of a gas at a given temperature is inversely proportional to the pressure to which it is subjected; the greater the pressure, the smaller the volume. At the time he formulated this law, he did not understand all its implications. Nearly two hundred years would pass before the relationship between pressure and volume that Boyle described became the cutting edge of cold research.

Most later appraisals of Boyle's life ignore his research on cold, though his contemporaries deemed it important, and it remains the first extensive scientific examination of the subject. Sensing that people might wonder why he had spent several years working on the cold, Boyle cited as his guiding rationale Bacon's identification of heat and cold as the right and left hands of nature. Expressing regret that cold had been "almost totally neglected" by classic authors, he also exulted because that neglect provided him with an "invitation [to] repair the omissions of mankind's curiosity toward a subject so considerable." He did so, splendidly, in his 1665 book *New Experiments and Observations Touching Cold, Or, An Experimental History of Cold Begun, To Which Are Added, An Examen of Antiperistasis, and An Examen of Mr. Hobs's Doctrine About Cold.* Many of Boyle's experiments had been conducted during the extremely frigid winter of 1662, but — the publisher John Crook wrote in a note to readers — the transcriber absconded to Africa and part of the original manuscript had been lost, forcing Boyle to redo some of the experiments and delaying publication. Crook noted he was rushing the book into print for the winter of 1665, so others would have the proper climate in which to repeat Boyle's experiments, should they choose to do so. In Boyle's own introduction, he likened exploring the cold to a physician attempting to do his work in a remote country where there was little help from implements or drugs. Reaching for another analogy, he wrote that the conditions in the far country of the cold might seem "incredible" to readers, as they had for him before he recalled that no one in such a very warm location as the African Congo was able to believe in the existence of ice. Boyle asserted he had "never handled any part of natural philosophy that was so troublesome and full of hardships" as the study of cold. He confessed to having suffered while conducting the experiments, but he reminded readers that sea divers "suffer as much wet and cold, and dive as deep, to fetch up Sponges, as to fetch up Pearls."

Boyle took as his task the establishment of basic information about the causes and effects of cold. Today it is difficult to imagine just how ignorant people were, as late as the latter part of the seventeenth century, of how ordinary cold operates. But the refusal of even some of the best minds of that century to accept Galileo's evidence about the earth revolving around the sun meant there was good reason for Boyle to attack what he called myths and misconceptions about the cold.

To dispel wrong-headed beliefs, Boyle researched every aspect of cold that any reader might wonder about: how one material might transmit cold to another; how atmospheric pressure related to cold; how cold condensed liquids such as oil; how salt, nitre, alum, vitriol (iron sulfate), and sal ammoniac (salt of ammonia) could intensify the cold; how cold could separate chemical solutions into crystals or salts. He made hundreds of experiments to eradicate confusion about the sources of cold, confusion he traced back to two of Aristotle's notions: that observation by the senses of unadorned nature was enough to apprehend the world, and that the source of all cooling in the world was a *primum frigidum.*

Had he believed Aristotle's contention that nature abhorred a vacuum, Boyle wrote, he would never have bothered trying to make one; thereafter, he refused to accept any ancient teachings without skepticism. To lead readers to distrust the Aristotelian adage about the evidence of the senses, in his book on cold Boyle reminded readers that tepid water flowing over a heated hand feels cool but actually is not very cold. He also took measurements in all seasons of a certain lake that many Englishmen swore was cooler in summer than in winter — because it certainly seemed refreshingly cool when one swam in it on a hot day — and showed that the lake's temperature was definitely lower in winter than in summer. Where Bacon had tried to refute Aristotle by philosophic opposition to his theories, Boyle added the evidence of experiment to reach the conclusion that the "testimony of our senses easily and much delude[s] us."

The ancients could not agree on the character of the *primum frigidum,* so Boyle tested their guesses. Aristotle had opted for water as the source of cold; this Boyle refuted by showing that substances with no water content, such as gold, silver, and crystal, could become quite cold. He also reported the observations of correspondents that ice forms atop the sea, where it interacts with air, but not at the sea's bottom; to Boyle this meant the *primum frigidum* was unlikely to be water. While on the subject, Boyle disposed of another Aristotelian-based contention, a theory called antiperistasis. Aristotle wrote that heated water cools more quickly than cold water; this was true, but only for previously boiled water, and only if the temperature at which one began to chill the water was not too high. More recent Aristotelians, without testing the limitations of the data, had pyramided the hot-begets-cold-quicker notion into an elaborate construct about the general action of opposites in spiritual as well as physical matters. Boyle exploded this rather silly theory by setting outdoors several vessels, each filled with water at a different temperature — hot, tepid, or cold — and collecting data showing that the rate at which the contents froze was not affected by the temperature of the water at the start.

With the same clarity of argument, Boyle dismissed two other candidates for *primum frigidum*. Plutarch had said it was the earth; but Boyle pointed out that the earth was known to be cooler and more solid near the surface and that other explanations had shown there was likely to be a central fire (not a central cold) at the core of the earth. Pierre Gassendi, a contemporary philosopher, suggested nitre as the *primum frigidum;* Boyle rejected this idea by pointing out that many cold substances exist without nitre or its "exhalations."

The "peripateticks," a group of scholastic natural philosophers, espoused air as the fourth candidate for *primum frigidum*. To refute that notion, Boyle pointed out that his correspondents had found ice in the middle depths of the sea, between top and bottom, which seemed to preclude air as the source of cold; moreover, he reminded readers, in his famous vacuum jar he had frozen water in the absence of all air. Boyle concluded that neither air nor earth nor nitre nor water could be the principal cause of cold, and that a *primum frigidum* was "but an unwarrantable conceit."

Now Boyle applied his ingenuity and pushed into virgin territory in a series of elegantly simple experiments. Many people had observed that when water barrels hooped with iron were left out in winter, they froze and the hoops broke. Various explanations had been advanced; Boyle thought them all nebulous, and that the only logical one was that the breakage was produced by the expansive power exerted by water as it froze. But how could he prove that his own guess was nearest to the truth? And kill two philosophic birds with a single stone? Another of Aristotle's theories was that substances were what nature had intended them to be and could not change; this led Aristotle to assert that when the form of a substance was altered — for example, when water became ice — that substance did not and could not gain or lose weight or size. A second assertion, from those who held that the *primum frigidum* was air, was that when a glass tube containing water was left outside overnight, and the next morning showed ice on its external surface, that ice must have been formed by cold air permeating through the glass from inside to out.

In a way few experimenters had done before, Boyle put down on paper the exact progression of his thoughts, so readers could follow his path to imagining the experiments he devised to refute outmoded contentions. His test to reveal whether anything had "migrated" from inside to outside was simplicity itself: he weighed the amount of water he put into a glass before letting it sit in the cold overnight, and the next morning again weighed the contents. Finding that the weight of the contents was the same in the morning as it had been before he put the vessel into the cold night air, he could then conclude that nothing had migrated from inside the vessel to cause the frost on its exterior. Boyle believed it most probable — he was very cautious in expressing his degree of certainty — that the ice that formed inside the glass overnight caused the glass to become cold and to foster condensation of water vapor on the exterior of the glass, which soon froze into the coating of frost.

Only after the preamble of that experiment could he proceed to the main task of demonstrating that freezing water did, indeed, change in size when it became ice. The least complicated way to show this would have been to measure the volume of water in a container before freezing, and compare it with the volume of ice after it froze. But Boyle knew that if he did not carefully control the conditions, the proponents of various candidates for *primum frigidum* would say any increase in the volume of ice was caused by migrating air, or by migrating particles from the iron or pottery or wood of the vessel's walls. So he decided to moot those possibilities by using a vessel made of glass, putting water in it, and then sealing the vessel. In other experiments, he had placed vessels outside to freeze, but to do so in this instance, he reasoned, might cause the glass to break. To prevent that, he worked inside a house, where he immersed the bottom of the vessel in a mixture of snow and salt. This ensured that the freezing of the water inside the vessel would proceed from bottom to top — allowing him to stop the process before the expansion had gone too far. Contending that his rigorous conditions had eliminated all other explanations, Boyle was able to convincingly conclude that ice was nothing more and nothing other than an expanded state of water.

But what were the parameters of that expansion? How greatly did the water expand in becoming ice? What amount of force did that entail? Aristotle and his followers had not even asked these questions, since they did not believe water expanded when it became ice. "No body has yet, that we know of, made any particular trials on purpose to make discoveries in this matter," Boyle noted. He ingeniously froze water in pottery and metal vessels with weights placed on top to retard expansion; he was astounded to discover that a weight of 74 pounds was required to prevent expanding ice from pushing out a cork. These results allowed Boyle to counter other faulty explanations of the action of cold. René Descartes, the "master of first principles" whose theories were very popular just after his death in 1650, contended that cold was only the absence of heat, which he defined as a free-floating "ethereal" substance that had neither weight nor mass; Cartesians believed cold was caused — in Boyle's words — by "the recess [receding] of that ethereal substance, which agitated the little eel-like particles of the water." Boyle wondered dryly how the ethereal particles could be so strong in their "recess" as to expand the water "with so stupendous a force." Epicureans held a related explanation: "cold corpuscles" that worked — again in Boyle's words — by "stealing insensibly into the liquors they insinuate themselves into, without any shew of boisterousness or violence." Were the vessels to be permeated in so calm a manner, Boyle observed, ice would never break them. There were, he concluded triumphantly, no "swarms of frigorifick atoms," nor was there any other explanation, other than his own, that could adequately describe or predict the phenomena he observed in his myriad experiments on the power of cold.

Boyle's most effective antagonist was Thomas Hobbes, who by midcentury had become one of England's leading philosophers. The disagreement between Hobbes and Boyle was, in part, that between a former "amanuensis" of Francis Bacon's and Bacon's leading scientific disciple. Hobbes championed Bacon's emphasis on the need to discover axioms and construct a comprehensive natural philosophy, while Boyle and his fellow members of the Royal Society more faithfully adhered to Bacon's insistence on properly conducted experiments, including deliberate attempts to manipulate nature. But the antagonism between Boyle and Hobbes was even more fundamental, involving two antithetical ways of seeing the world and of discovering facts about it. Following Aristotle more than Bacon, Hobbes's natural philosophy used deduction from "first principles," a process that replicated the path he had taken to reach his unusual understandings of civil law and ethics. In Hobbes's view, when a philosopher wanted to know something about the workings of the natural world, to begin with he made a "first principles" hypothesis, then inferred phenomena and rules from it, and from these drew his conclusions; in accordance with this way of determining knowledge, experimentation had no value, because it could not reveal anything about the world that the first principles had not already predicted.

This was directly counter to Boyle's experimentalist way of viewing, questioning, and assessing the world; both he and Hobbes knew how sharply opposed their viewpoints were, and in a series of books published over several decades, they directly attacked one another's contentions, methodology, and conclusions. In *Dialogus Physicus* and in *De Corpore,* Hobbes mounted a strong critique of experimentation and of Boyle's use of it in regard to the vacuum pump. He contended that experiments were unable to "establish" facts. As an example, he noted that Boyle's air pump had leaked, so whatever results it produced were faulty and could not be the basis for making explanations. Moreover, Hobbes argued, for the results obtained, there were alternative causal explanations to those advanced by Boyle.

Stung by the criticism of his famous device and of his experimental method, Boyle took steps to shore up both, repairing some of the pump's inadequacies before his next series of experiments on atmospheric pressure, and inserting into his subsequent definitions of the experimental method the need to generate a hypothesis consistent with all known facts about nature — a hypothesis that could explain not only the results of the experiments at hand but also of all similar experiments, and of future tests not yet conceived. These were salutary results of Hobbes's critique, in that they forced the leading English practitioner of the experimental method toward greater rigor of methodology.

Hobbes left himself vulnerable to significant counterattack from Boyle by venturing opinions in an area about which Boyle knew far more than Hobbes: the cold. Boyle began his riposte by claiming an obligation to respond because Hobbes's "fame and confident way of writing might prejudice experimental philosophy in the minds of those who are yet strangers to it," and who might "mistake confidence for evidence." According to Boyle, Hobbes's theory about the origin and actions of cold was "so inconsiderately pitched upon, and so slightly made out" that it should not merit more than a passing mention, but since it was popular, it had to be refuted. Boyle cited Hobbes's discussion of how ice forms, in which Hobbes wrote that the source of all cold was wind, which

> rakes the superficies of the earth, and that with a motion so much the stronger, by how much the parallel circles towards the poles grow less and less. From whence must arise a wind, which will force together the uppermost parts of the water, and withal raise them a little, weakening their endeavour towards the center of the earth.

Citing such convoluted reasoning, Boyle noted that Hobbes gave no proofs or demonstrations of his theories, only explanations that were "partly precarious, partly insufficient, and partly scarce intelligible."

Hobbes's contention that cold winds made the exterior parts of a body coagulate and go inward, thus transmitting the cold, was wrong in almost every detail, Boyle wrote. Boyle had put live animals in his vacuum apparatus, extracted the air, and then frozen the animals in the absence of all air — which disproved Hobbes's thesis that winds were what cause bodies to feel cold. He also showed that when a cake of ice served as a stopper in a vessel between the wind and the remaining water inside, freezing continued despite the absence of wind. There was no way, Boyle concluded, that Hobbes's theory of wind-as-source-of-cold could explain Boyle's experimental results.

It was with such counterattacks, in this and other areas of physics and chemistry, that Boyle eclipsed Hobbes in natural philosophy and relegated Hobbes's contributions in science to the dustbin of history. Hobbes never became a Fellow of the Royal Society — blackballed, it appears, by Boyle and a few others — though he did remain on good terms with many of the Royal Society's early members. Within a few years of his death in 1679, at the age of ninety, his reputation had come to rest mainly on his contributions to political philosophy, ethics, and morals, for which he is still principally known today.

In 1665 the effacement of Hobbes by Boyle in natural philosophy lay in the future, and Boyle used Hobbes's arguments as a sort of governor on his own exuberance. This was most evident in a late chapter of the book on the cold, a "sceptical dialogue" involving the fictional character Carneades, who was a stand-in for Boyle. In it, Carneades agreed with Bacon that cold "must be a privation [deprivation]" of motion of some sort but admitted he could not demonstrate precisely how it worked, and he could not completely refute all other explanations, such as that cold might be transmitted through the walls of vessels, in the manner of rays of light. Boyle was in effect conceding Hobbes's point that there could be alternative explanations for certain of his experimental results. But if Boyle could not definitively assert that his theory about the deprivation of motion as the cause of cold was the only correct explanation, he could and did contend that he had succeeded in disproving every other explanation, and that his had reached Huygens's threshold of the fairly probable, which was the best that could be achieved at that moment.

"Future industry," Boyle predicted, would be able to build on his work, venturing beyond the frontiers to chart and explain to the country of the cold. In that future exploration, its Columbus wrote, if any one thing was needed more than another, it was a good and reliable thermometer, the lack of which had forced Boyle to leave some important experiments on cold "untried."

Battle of the
Thermometers

THE CAPSULE VERSION OF SCIENCE HISTORY holds that in a stroke of genius, Galileo invented the thermometer in 1592. The real story is more complicated. In northern Italy just then, people were touting the wonders of a J-shaped tube, closed at one end and filled with water that rose and fell during the day like the tides of the sea, the movement supposedly influenced by the moon. A *scherzo*, a trick, Galileo fumed, and set out to show what actually moved the water — rising and falling temperature. Half-filling a narrow-necked glass flask with colored water, he suspended it upside down in a bowl of more colored water; when the temperature went up or down, the air contained in the bulb expanded or contracted, moving the column of water in the neck down or up.

Galileo's device was not a thermometer — it was a thermoscope, which records the presence of heat but lacks a scale to measure relative heat. Moreover, he may not have designed the thermoscope on his own but instead appropriated a device from Santorio, a colleague and professor of medicine at Padua. Both men may have been attempting to reproduce a demonstration made by Hero of Alexandria in the first century A.D., itself modeled on work by Philo of

Byzantium in the third century B.C. Documents show that Galileo had read a 1589 Italian translation of Hero's *Pneumatics.*

A great deal more about heat and cold awaited discovery, and as Bacon wrote, "It would be the height of folly — and self-defeating — to think that things never heretofore done can be accomplished without means never heretofore tried." In this history the artificer has had his day; and the natural philosopher; and the dogged experimentalist. Now it was time for the instrument maker to provide the "means never heretofore tried," equipment to enable would-be explorers of the far extremes of temperature to travel further, learn more, and accomplish "things never heretofore done."

Seventeenth-century natural philosophers had mathematical dreams. Those who yearned to better understand the heat and cold wanted to subject them to mathematical analysis. What was the relationship between the temperatures of melting ice and of boiling water? Precisely how much colder was ice than snow? How hot did dry wood have to become before it burst into flame?

Thermoscopes could not answer these questions, but an attached mathematical scale might give some answers, depending on the reliability of the power moving an indicator up and down the scale. Friends touted to Galileo the 1598 *perpetuum mobile* of Cornelis Drebbel, believed to rely for its motion on the expansion and contraction of air. Galileo's first true thermometer, made in the first decade of the seventeenth century, used heated air to move the indicator, a fact that made some later writers ascribe the invention of the thermometer to Drebbel. There is not much to that claim, since other contemporary thermometers also used heated air. A stronger attribution for the invention of the thermometer can be made for Santorio. He published an important commentary that reproduced the attempt by the Greek physician Galen in the second century A.D. to measure heat and cold along a scale, and he provided his own design for a thermometer to the instrument maker Sagredo, who constructed several of them and then wrote excitedly to his teacher Galileo about using them to discover "marvelous things, as, for example, that in winter the air may be colder than ice or snow; that the water just now appears colder than the air; that small bodies of water are colder than large ones."

Sagredo's note to Galileo confirms that the near borders of the country of the cold had never before been accurately mapped, because prior to this moment, no one could prove that winter air was physically colder than ice or snow.

Although these first thermometers did have a scale, to glean more useful information scientists required a better scale, one whose divisions denoted intervals that had meaning and that showed the temperature in relation to one or more fixed reference points. What intervals should be used? What points could be designated as fixed? If there was a zero, where in the scale should it be put? And what would that zero signify? Would thermometer A in location B always give readings comparable to those obtained from thermometer C in location D?

Those questions had to wait for answers until the solution of the motive-power problem. In the 1640s, when Otto von Guericke in Germany, Boyle in England, and Evangelista Torricelli in Italy proved that air pressure varies with a location's height above sea level and with changing weather conditions, the use of air as a motive power in an open thermometer had to be abandoned. Grand Duke Ferdinand II and his Accademia del Cimento in Florence took up the problem of what motive power to substitute for open air in a thermometer.

Like the Royal Society in London and the Académie des Sciences in Paris, the Accademia del Cimento was founded by distinguished experimenters and had highborn patronage. The Accademia lasted a mere ten years, though, and perhaps because of its short life it has not been invested with the same reverence by later generations — but it was crucial to the history of thermometers. It was an institution devoted to experimentation, as evidenced by its name, translated as the "academy of the concrete" or the "academy of experiments," and by its motto, *Provando e Riprovando,* "proving and proving again." The father of Ferdinand II and of Leopold Medici had been a student of Galileo's, and the sons were fascinated by the work of Torricelli; those ideals translated into their installing facilities for the use of ten scientists in the Pitti Palace in Florence in the late 1650s. The Medicis' Accademia flourished at a moment when the imprimatur of a scientific institution could have the greatest influence on the exploration of the cold, by influencing the acceptance of its critical tools over those produced by other groups.

In contrast to the English and French patrons of science, the Medicis were actively involved in the institution's experimental work. Ferdinand was a corpulent man with a bulbous nose and a black mustache whose ends rose up toward his eyes. "The grand duke is affable with all men, easily moved to laughter and ready with a jest," a contemporary account stated. One of his heartaches was his inability to interest his son in scientific work; the "melancholy" teenager exhibited "the symptoms of a singular piety" that dismissed experimentation as incompatible with religious faith. Ferdinand's brother Leopold was quite devout, but that did not prevent him from being a serious scientist. He spent four hours a day reading books on literature, geography, science, architecture, and general curiosities; it was said that, like a little boy with a piece of bread, Leopold always kept "a book in his pocket to chew on whenever he [had] a moment to spare."

The Accademia shared with the Royal Society the goal of discrediting the pseudoscience of Aristotle's followers. The Medicis corresponded with the Royal Society, and they employed a secretary to carefully note the Accademia's experiments on the incompressibility of water, the gravity of bodies, and the electrical properties of metals and liquids. But mostly the men in the Pitti Palace addressed the task of making more capable and subtler measuring implements of every sort then in use — barometers, hygrometers, telescopes (Ferdinand ground his own lenses), astrolabes, quadrants, calorimeters, microscopes, magnetic devices, and thermometers. One visitor reported seeing "weather glasses" (barometers or thermometers) even in the grand duke's bedchamber, and another marveled at thermometers displayed as works of art rather than as implements of science. There were thermometers to measure the air temperature, thermometers to measure heat and cold in liquids, thermometers for baths. During the winter, *strumentini,* little temperature-measuring thermometric instruments, hung in every room of the palace.

Several members of Ferdinand II's court were, in the words of a food historian, "ice-mad," among them the secretary, who reverently called snow "the fifth element" and wrote novels in which characters begged one another for ice-based treats. Other courtiers were known to delight in the ice goblets and fruit bowls, the ice pyramids, and the table-sized ice-capped mountains of the Medici celebratory feasts.

It was Ferdinand who invented the sealed-glass thermometer. When the Medici workshop devised one with a scale that marked off into fifty segments the bulb in which the measuring fluid was held, Sagredo accepted that innovation and used it as a base from which to further divide his new thermometers. On these, he marked off 360 divisions, like the gradation of the circle; after that, all scientists started calling the divisions "degrees."

Most of the Florentine sealed-glass thermometers used "spirit of wine," distilled alcohol. Small glass bubbles filled with air at varying pressures hovered within the liquid, changing position as the temperature rose or fell. Later the Accademia flirted with the use of mercury, but its scientists soon — and for reasons that to this day remain obscure — abandoned this choice and went back to spirit of wine. Spirit thermometers did not work well at the low and high ends of the scale, because alcohol boils at a lower temperature than water, and it freezes lower, too; moreover, because the density of the distillate varied from batch to batch, spirit thermometers were often incompatible with one another. Counterbalancing such inadequacies was the high quality of the thermometers produced by the

Accademia del Cimento, which was why those that bore its good name came to be requested and used throughout Europe.

Just when the Accademia had built up enough expertise to be well on the way to solving the remaining technical problems with thermometers, quarreling among the members reached an absurd height. In 1657 one member had written that "only disorder is to be found" at the Accademia. By 1666 some members could only be induced to speak at meetings if certain others were absent. The institution decayed. Louis the Sun King lured Christiaan Huygens to become the director of his Académie Royale and enlisted one of the key members of the Accademia del Cimento with a promise of a pension and the right to publish his experimental results under his own name; two others defected elsewhere, ostensibly to seek better climates for their health.

Leopold was discouraged by the infighting, which was going on at the same time the Catholic Church was requesting that one of the Medicis become a cardinal. A former Accademia member's memoir suggests that shutting down the Accademia was made a condition for awarding Leopold the red hat of a cardinal, though other contemporary accounts dispute this notion. In any event, there was an abrupt end to the diary of the Accademia's experiments at the time Leopold was summoned to Rome. He sent a passing-the-baton letter to Constantijn Huygens, entreating him "to look over the great book of nature by means of experiments, and find new things never heard of before, and to purge books of those experimental errors that have been too easily believed, even by the most esteemed authors." In March 1668 Leopold journeyed from Florence to Rome to become a prince of the church; in the coach, as though in defiance of leaving science behind, he took along one of the Accademia's last *strumentini* and whiled away the hours of his journey recording its changing observations.

Boyle obtained a Florentine thermometer, as did other English experimenters; Hooke tried to improve Boyle's in several ways, not the least of which was to substitute mercury as the liquid within the glass tubing; once again, however, for some unknown reason, mercury was soon rejected in favor of spirit of wine. Hooke adapted Florentines for himself and for his friend Christopher Wren. There was even an attempt to put the Royal Society's imprimatur on a thermometer adapted from a Florentine model with Hooke's modifications. Political considerations quashed that, but Hooke's own thermometric innovations advanced the field.

By the 1660s this son of a parish curate, though still in his twenties, had matured from his stint as Boyle's assistant to become one of the most inventive and mechanically able of the Royal Society Fellows, having done significant work in microscopy, astronomy, geology, combustion, and meteorology. He also helped Wren rebuild London after the Great Fire of 1666. When the diarist Samuel Pepys became a member of the Royal Society, its president told him that Hooke did the most, and promised the least, of any of the Fellows. Bent almost completely askew, Hooke was described as "meanly ugly, very pale and lean"; John Aubrey agreed with that description but balanced the picture by adding that Hooke was "of great suavity and goodness." In later life, Hooke would run afoul first of Isaac Newton, then of Henry Oldenburg, secretary of the Royal Society; their combined opposition would succeed in downplaying his accomplishments and undercutting the acceptance of his work by the wider world.

Those accomplishments were considerable, especially in regard to the scientific history of the cold. Hooke was the first person to minutely grade a thermometric scale with marks representing a precise volumetric quantity, each equal to one-thousandth of the expanding volume of spirit in the bulb. He was also among the first to assert that thermal expansion might be a general attribute of matter. "This property of Expansion with Heat, and Contraction with Cold, is not peculiar to Liquors only, but to all kinds of solid Bodies also, especially Metals," he wrote, and italicized his conclusion: "*Heat is a property of a body arising from the motion or agitation of its parts."* Two hundred years would elapse before James Joule would prove this same conclusion experimentally and deduce from it the idea — central to the further exploitation of the cold — that heat is a form of energy related to the motion of atomic particles. Hooke was also among the first to propose establishing a permanent flagpost in the country of the cold, one that could be used for navigation by all later explorers: he sought to make the freezing point of water a fixed point of reference on a thermometer.

Today the need for a thermometer to have such a fixed point may seem obvious, but near the end of the seventeenth century, fixed points were a matter of controversy. Some people thought the freezing point of water varied with the time of day, the latitude, or the season. Following the tenets of Hobbes — the old notion that atmospheric conditions changed as one approached the planet's poles — even such usually astute men as the English astronomer Edmund Halley contended that the freezing point of water would not be the same in London as in Paris. Not until the 1730s was that notion finally disproved.*

Among the candidates for fixed points proposed during the second half of the seventeenth century were, on the lower end, the temperature of mixtures of salt and ice, of ice and water, and of near-freezing water. In the middle range the suggested fixed points were the temperature of the deepest cellar of the Paris Observatory (believed not to vary between summer and winter); the congealing points of aniseed oil, linseed oil, and olive oil; the temperature at which butter melts; and that at which wax melts. On the higher end they were the heat of the healthy human body, measured under the arm or in the anus; the internal temperature of certain animals; the maximum summer temperature in Italy, Syria, and Senegal; the boiling points of pure alcohol, spirit of wine, and water; the heat of

* The freezing point of water did become an acknowledged fixed point in the minds of scientists and instrument makers for many years, until it was determined in the nineteenth century that the melting point of ice, which is slightly different, was a more precise reference location.

a kitchen fire hot enough to roast foods; and the presumed temperature of the sun's rays. Halley, another Fellow of the Royal Society, recommended the boiling point of spirit, yet he also revealed why it was so difficult to establish any fixed point: "The spirit of wine used to this purpose [must] be highly Rectified or Dephlegmed for otherwise the differing goodness of the spirit will occasion it to boil sooner or later, and thereby pervert the designed exactness."

The questions of fixed reference points and measuring liquids even lured Isaac Newton. An eighteenth-century historian of thermometers commented that in the way Newton "carried everything he meddled with beyond what anybody had done before him, and generally with a greater than ordinary exactness and precision, so he laid down a method of adjusting thermometers in a more definite way than had been done hitherto."

Newton forthrightly set his zero at the freezing point of water, but his indication of the exact location of that zero reveals the difficulty of being precise as to where a flag should be planted: he described the point as "[t]he heat of the air in winter, when the water begins to freeze; and it is discovered exactly by placing the thermometer in compressed snow, when it begins to thaw." Newton was wrong in many of his thermometric ideas. His scale erred in being self-referential — the meaning of its points could be understood only by reference to one another, and not by reference to any widely recognized standards — and in having a bias toward measuring the upper temperatures more finely than the lower. As Newton went up the scale, his descriptions became longer and more exact. Number 17 was "Greatest degree of heat of a bath, which a man can bear for some time without stirring his hand in it," but above that, his choices increasingly involved materials that ordinary people would seldom encounter, such as the boiling of mixtures of water and metals, and they also used ratios that smacked of the magical — for instance, he insisted that the heat of boiling water was three times that of the human body, six times that of melting tin, and eight times that of melting lead, and that the heat of a kitchen fire of heaped coals was sixteen or seventeen times that of the human body.

Newton contributed in a more novel and useful way in the matter of choosing the range of information the instrument could display, by proposing linseed oil as the fluid inside the thermometer bulb. This extract of flax was a brilliant choice, because it addressed the basic requirement that thermometers be able to measure accurately on the high and low ends of the scale. Being extremely viscous, linseed oil could remain liquid at temperatures above that at which water boiled and below that at which water froze. But linseed oil's viscosity made it slow to register changes — when the temperature was raised just a few moderate degrees, linseed oil would drain in a sluggish way down the sides of the tube on which the scale was marked. This long adjustment time made the linseed-oil thermometer difficult to use and may have been why a 1701 article on it by Newton in the Royal Society's *Transactions* was unsigned and only later attributed to him.

The next innovator in thermometry was French physicist Guillaume Amontons, regarded as a relatively obscure figure today, but one whose work provided a critical foundation for others in the exploration of the cold, even well into the twentieth century. The son of a provincial lawyer, Amontons was deaf since birth, and he had been largely self-taught in the sciences when in 1687 he first sought attention for his work from the Académie Royale. The twenty-four-year-old demonstrated a new hygrometer that used two liquids for measurement, one of them mercury, and followed this up a year later with ideas for multiliquid barometers and thermometers. More than the innovations of a craftsman, these were the constructions of a man who understood basic principles and how to apply them. The instrument maker for the Paris Observatory thought Amontons's ideas worthy enough to be discussed with members of the Royal Society on a visit to London. Put into a 1695 book, these ideas earned Amontons admission to the Académie.

His main contributions to thermometry and to the study of cold came by way of a detour. Amontons tried to create a "fire-wheel" that used the heat of a fire to expand air and make it move a wheel; he failed, but the principles he elucidated were sound and would be built on in later research by others on heat engines, horsepower, and friction. He also used some of the fire-wheel research to create better thermometers. Heating three unequal masses of air and water in glass bulbs submerged in boiling water, he demonstrated that the masses "increase[d] equally the force of their spring by equal degrees of heat," causing the air pressure to rise by one-third of an atmosphere in each bulb — 1 atmosphere being equal to the pressure of air at sea level, 14.7 pounds per square inch.

From this pressure work Amontons drew two important conclusions. First, even if the heated air had been afforded "the liberty of extending itself" instead of remaining confined within the glass bulbs, it still would not have increased its volume by more than one-third.* Second, since the water in all three bulbs boiled at the same temperature, despite variations in the volume of the water and air in the bulbs, the boiling point of water was proved to be a constant, one that could be used with confidence as a fixed point on a thermometer. Based on these results, Amontons designed a new air-based thermometer employing sealed glass; the sealing prevented distortions in the readings that would otherwise come from changes in atmospheric pressure.

Amontons also advanced the cause of better thermometry by his tough criticism of that anonymous 1701 paper on the linseed-oil thermometer in the *Transactions* of the Royal Society — not knowing it had been written by Newton. He lambasted the author's contention that a body that on the author's self-referential scale had a

---

* Another hundred years would elapse before French physicist and chemist Joseph-Louis Gay-Lussac would definitively refine this notion into the law that at constant pressure, all gases have the same coefficient of expansion.

temperature of 64 was twice as hot as one that registered at 32, charging that not enough was currently known about heat and cold to support such an assumption.

As his critique of Newton's thermometry suggests, Amontons was a purist who did not often engage in the speculation rampant among researchers of the era. But he was a man who did pay attention to the implications of his mathematical computations; so, having proved that air contracts by a fixed proportional amount when cooled, he could not avoid calculating what *might* happen to air if its temperature were radically reduced, to well below the freezing point of water. Would air become denser, and as the temperature dropped further, would air become a liquid? Would that liquid be water — current thinking favored that conclusion — or something else? In the total absence of heat, would there be any air pressure? In a 1703 paper, Amontons evolved a simple equation showing that a total absence of heat was theoretically possible.

In the equation, the product of pressure times volume equals the product of temperature times an unknown constant. From this Amontons drew the clear implication that if by some means the product of pressure times volume became zero, on the other side of the equation the temperature could fall to an "absolute zero." Amontons did not come right out and say there was an absolute zero, because he considered such a thing incompatible with what was currently known about nature and with what he believed.

For Amontons, absolute zero was a hypothetical construct to be imagined, not to be realistically pursued. Two years after his article was published, he died of an internal inflammation, at the age of forty-two. Central to his legacy was his understanding that in the grand scheme of things, human beings and most other life on earth lived not far, in temperature terms, from the freezing point of water, and that the country of the cold that began at the freezing point was far more vast than human beings had previously believed, promising temperatures below what any scale then in existence could measure, down to an almost mythical point, an absolute zero, the end of the end.

Around 1702, while Amontons was doing his best work in Paris, in Copenhagen the astronomer Ole Rømer, who had calculated the finite speed of light, broke his leg. Confined to his home for some time, he took the opportunity of forced idleness to produce a thermometer having two fixed points, marking his scale at 7.5 for the melting point of ice and at 60 for the boiling point of water. His zero was thus well below the freezing mark, and supposedly represented the temperature of a mixture of salt and ice, while blood heat on this scale happened to fall at 22.5, or three times the melting-ice temperature. Rømer wasn't concerned very much with the upper and lower limits of his scale, because his primary work was in meteorology, which dealt with temperatures in the middle range. Six years after his broken-leg episode, Gabriel Daniel Fahrenheit visited him in Copenhagen. The young, Polish-born man had become fascinated by the making of scientific instruments and wanted tips on the techniques involved. And he may also have had another important reason for visiting Rømer.

Born in 1686 in Danzig, Fahrenheit had been orphaned when he was fifteen by the sudden death of both parents from mushroom poisoning on a single day in 1701. He had then been sent by his guardians to Holland as an apprentice to a bookkeeping firm, but he had run away from that position so many times that his guardians had had a warrant put out for his arrest. Going from city to city, he developed an interest in scientific instruments, teaching himself by visiting laboratories, refusing to settle anywhere, perhaps out of fear of apprehension and of being returned to his apprenticeship. What Rømer may have done for Fahrenheit, in addition to exciting his interest in thermometers, was arrange for the withdrawal of the arrest warrant, a courtesy that Dutch authorities would have extended to Rømer as the mayor of Copenhagen.

Until 1716, as Fahrenheit traveled from place to place, he worked on his thermometers, solving technical problems and becoming a competent glass blower. For several years, he sought the patronage of German philosopher and mathematician Gottfried Wilhelm von Leibniz, and it was only after Leibniz's death that Fahrenheit decided to settle permanently in the Netherlands and begin cultivating the patronage of Hermann Boerhaave, the celebrated physician, chemist, and botanist. (Boerhaave's fame was so great that a letter sent from China reached him even though the address read only "Boerhaave, Europe.")

Fahrenheit sent to Boerhaave and to other leading scientists samples of his thermometers, and he sought commissions to make more. Boerhaave commissioned several and also requested that Fahrenheit do some experiments for him. In these, Fahrenheit established that the boiling point of water is always a function of the atmospheric pressure and that for each atmospheric pressure the boiling point of water is fixed. He drew from these facts the implication that height and depth could be measured by a thermometer, so long as it was able to accurately record the point at which water began to boil. Boerhaave reported Fahrenheit's experiments in his influential chemistry textbook *Elementa Chemiae*.

To make his thermometers, Fahrenheit adopted but also importantly altered Rømer's scale. Finding the numerical markers Rømer had used for melting ice and blood heat, 7.5 and 22.5, "inconvenient and inelegant on account of the fractional numbers," he tried to set his zero lower by making his blood-heat mark higher, at 24; this put his melting-ice temperature at 8. He made some early thermometers with this scale, but the aesthetics of these numbers still did not satisfy him, and he also wanted a scale on which each degree reflected a fixed percentage change in a liquid — as did the devices made by Boyle, Newton, and Hooke. With inexact implements, he nonetheless managed to create a scale on which a 1-degree rise or fall produced a change of one five-hundredth the initial volume of spirit of wine at the zero point.

There his innovations might have ended had Fahrenheit not been unusually inquisitive, willing to master new languages to advance his knowledge — French, so he could read Amontons and other contributors to the *Mémoires de l'Académie Royale,* and English, so he could read Boyle, Newton, and Hooke in their native tongue. After reading Amontons, Fahrenheit switched to using mercury and recalibrated his scale. On his newer thermometers, each degree corresponded to one ten-thousandth the initial volume of the mercury, the same proportion as in Boyle's and Newton's thermometers. To achieve that ratio, he had to quadruple the values on his scale — which turned out well, because with the melting-ice point at 32° and the blood-heat point at 96°, he now had a scale on which the key numbers were still divisible by 4, but on which the range was greater, making it easy to work with. The ability to more accurately locate blood heat was quite important, because Fahrenheit knew of Boerhaave's interest in measuring body temperature and wanted his influential patron to adopt (and recommend) his devices.

Perhaps to preserve his ability to exclusively manufacture these thermometers, Fahrenheit did not publish the calculations that had led to his scale. Knowles Middleton, the modern authority on the history of thermometers, suggests that all instrument makers concealed such matters, or obfuscated them, to prevent others from replicating their instruments without paying for them. Thus while Fahrenheit promised to provide Boerhaave with "accurate descriptions of all the thermometers which I make, and of the way in which I have . . . attempted to rid them of their defects, and by what means I have succeeded in doing so," he actually withheld the secret of the volumetric measurements that had helped him arrive at the important numbers of 0, 32, and 96. An unintended consequence of this concealment was that for hundreds of years afterward, scientific historians wrote that Fahrenheit's scale was arbitrary.

In 1724 Fahrenheit was elected as a foreign member of the Royal Society and went to London. Just before this trip, he did some significant though haphazard basic experiments to determine some fixed points for his thermometers. In a 1729 letter to Boerhaave,

Fahrenheit first apologized for any delay in delivery of his thermometer, saying he was "now seeking a friend of the fair sex, which matter, as you will understand, will take up much of my time until it is resolved," then told of his pre-1724 research on the artificial production of cold.

These experiments were like a chance visit to an unknown territory by someone unprepared to be a proper explorer but who grasped the opportunity of an accidental landing to climb a mountain, view the lay of the land, and record a few observations of unusual phenomena. Mixing aquafortis (concentrated nitric acid) and ice, Fahrenheit reduced the temperature of his mix to a point so low that the largest measuring tube he had was not long enough to measure it. Intrigued, he constructed a thermometer able to register as far down as 76 below his 0 and managed to lower the temperature of an aquafortis/ice mixture to 29 below.

Next, using a series of beer glasses with freezing mixtures in them, he tried what would later become known as a cascade series: he cooled the first glass with his original aquafortis/ice mixture until the mix registered its lowest temperature; then he poured off the liquid in the mixture and used the solid that remained on the bottom as the starter for the second glass, to which he added more liquid, enabling him to reduce the temperature of the second glass to 32 below; continuing the technique, he reached down successively to 37 and then to 40 below. He boasted to Boerhaave that he could have gone still lower had he had purer chemicals. During this exploratory descent, Fahrenheit also managed to create crystals from some liquids. Not well enough schooled to make advanced scientific conclusions from his experiments, he "humbly" wrote to Boerhaave the prescient observation that "we know just as little of the first commencement of heat as we know of the extreme limit of heat."

After his sole excursion into the territory of the cold, Fahrenheit returned to what he did best, the manufacture of measuring implements, and confided in his next letter to Boerhaave that friends of the young lady he was courting, who wanted her money, had influenced her to spurn his advances. Free of the distraction of finding a wife, he immersed himself in the technical problems affecting his thermometers' precision of measurement and comparability with one another. He was vexed because the various kinds of glass used for tubing yielded differing results and because the inconstant purity of the mercury he bought made it nearly impossible to accurately fill two bulbs with precisely the same amount of liquid mercury — he would tolerate only an error margin of 5 parts in 11,520, or 0.05 percent.

Although his thermometers were bought and used throughout Europe, Fahrenheit died penniless in 1736; he had spent everything he earned from the ones he sold on research and on materials for newer ones. Good as Fahrenheit's thermometers were, they came into general use only in those countries where the language of choice was not French. No matter what the field of endeavor — astronomy, agronomy, or meteorology — the French liked to use their own measurements; there were, for instance, 18 kinds of the unit of length known in French as the *aune,* and in one district, 110 measures for grain. The French considered Fahrenheit to be Dutch/Polish and would not use his thermometers. Instead, they preferred thermometers with a scale created by René-Antoine Ferchault de Réaumur. This highborn savant had done very credible work in botany, ferrous metallurgy, and embryology, and he considered his thermometric work less worthy of attention than his other scientific endeavors. He was right about that, for his thermometric scale was not even a true innovation, simply reproducing Hooke's scale without attribution. Yet Réaumur wrote so extensively about his thermometric scale that his scientific memoirs took on the force of public-relations efforts and pushed the French to adopt it, even though Réaumur thermometers froze at low temperatures and had to be modified to be useful. Réaumur also advanced his cause by actively dissuading experimenters from using Fahrenheit-scale thermometers, and when other thermometers marked with a "centigrade" scale began to show up on the Continent, he went out of his way to disparage and suppress them, too.

In this latter effort he failed, because in time the centigrade scale, the most logical and most useful measuring innovation developed during the first 150 years of the thermometer, would supplant Réaumur's and almost all other scales. In terms of accuracy, fixed points, and utility of the scale to the needs of researchers, centigrade-scale thermometers reached as far as was then technologically feasible. They became extremely important tools for exploration of low temperatures.

In 1948 the Ninth International Conference on Weights and Measures decided that the points on the centigrade scale that had been known as "degrees centigrade" should henceforth be known as "degrees Celsius," to honor Anders Celsius, who had invented the scale. Success has many fathers, and despite the twentieth-century canonization of Celsius as the inventor of centigrade, there remains a whiff of controversy as to the true claimant to that title.

In 1740, the year before Celsius was supposed to have invented the scale, Réaumur was grumbling to his diary about centigrade scales, and there is also evidence that several Swedes other than Celsius could lay claim to having formulated the centigrade scale. The instrument maker Daniel Ekström had worked in England with the thermometer maker for the Royal Society, and an Ekström-modified London thermometer appears to have been in use at Uppsala University since 1726 — a thermometer on which the freezing point of water was at 0° and the boiling point at 100°. Two other potential claimants for the title of father of centigrade were Mårten Strömer and the botanist known to history as Linnaeus.

Anders Celsius's centigrade scale had the distinction of being born as a direct result of a visit to the geographic country of the cold. Celsius was the son and grandson of astronomers, and his main work was in that discipline; he mapped the aurora borealis, studied the light of the moon, changed the Swedish calendar from the Roman Julian to that used by other countries in the eighteenth century, and was a participant in an international scientific expedition that journeyed far inside the Arctic Circle to help verify Newton's theory that the earth was flattened at the poles. On that trip in the 1730s, Celsius became dissatisfied with the instruments then available for measuring the considerable cold in the air and on the ground.

He was an unusual man, recalled by memoirs as being quite learned, able to converse in a half-dozen languages and interested in ancient runes, a man of "harmonious" personality and grand vision but also a man of honor, one who defended ideas he thought were correct, his own as well as those of Newton — this last contention reflecting a controversy in which Celsius's astronomical observations that verified Newton's were viciously attacked by a leading Continental scientist, then eventually proved correct.

In 1741 Celsius obtained a thermometer from St. Petersburg and etched on the side opposite its scale a formulation of his own. It had two fixed references, the boiling and freezing points of water, which he designated, respectively, as 0° and 100°; like his fellow astronomer Rømer, Celsius was more concerned with the accuracy of the near-freezing temperatures than with the near-boiling ones. His thermometer was first used on Christmas day 1741, and he wrote about that use in a 1742 publication, stressing the handiness for calculations of having 100 demarcations between his two fixed points. Two years after this publication, Celsius died, at the age of forty-two.

Strömer made a centigrade thermometer in 1750, on which he simply reversed Celsius's scale, putting 100 at the top and 0 at the bottom; as Celsius's successor at Uppsala, Professor Strömer would not think of claiming credit for his predecessor's innovation. But Carolus Linnaeus could and did try to do so. In 1758 Linnaeus claimed to have been the originator of the centigrade scale.

That Linnaeus would even want to make such a claim underlines the importance to science the centigrade scale had achieved in a relatively short period of being in use. In 1758 Linnaeus was the most celebrated botanist in the world, a man whose fame as a scientist outshone that of virtually all others of his time. He was also one of the most egocentric scientists who ever lived, a man who told others that God had anointed him to promulgate the rules of classification and who believed that all his publications were masterpieces. Linnaeus brooked no criticism of himself, even as he vacillated between extremes of exhilaration and despair over his work and worth. Knowles Middleton suggests that in Linnaeus's mind, his claim on centigrade probably dated back to 1735, when he had lived for a time on the estate of a wealthy Dutch planter and had written a book about the gardens there; one illustration shows angelic putti holding what appears to be a thermometer with a scale that goes from 100 at the top, down by tens to 1 in the middle — not to 0 — and then down again by tens to 100 at the bottom. However, Middleton points out, the text of Linnaeus's book does not refer to the instrument in the illustration at all but does mention a hothouse containing African violets kept at a temperature of 70. Since on a centigrade scale a 70° temperature would scorch such plants — 40° would be about their limit — it is likely that Linnaeus's text reference was to a thermometer etched with a non-centigrade measuring scale. So Linnaeus can be credited, at most, only with inverting Celsius's scale.

As for the date of his claim to inventing the centigrade scale, the botanist apparently chose not to initially lay claim to inventing the centigrade scale before 1758, because earlier he was indebted to Celsius, who was pushing hard to obtain something for Linnaeus that he desperately wanted at that time — a position at Uppsala University. In deference to his sponsor, Linnaeus delayed making a claim of invention of the centigrade thermometer scale until more than a decade after Celsius was dead.

Adventures in the Ice Trade

DURING THE EXCEPTIONALLY FRIGID WINTER of 1740, workers in St. Petersburg harvested from the Neva River square blocks of pure ice and with cranes and pulleys hauled them to the bank. Some blocks were so large that each took up an entire horse-drawn wagon, so it appeared to onlookers as though the wagon was burdened with one huge diamond or sapphire. The workmen were said to resemble mythical or allegorical figures, their beards and hair caked with congealing ice, their visages contorted by grimaces as they toiled in the bone-chilling cold. Using water to fasten one block to another, the workers erected, for the empress Anna, a translucent palace 56 feet long, 21 feet high, and 18 feet in depth. They painted the window frames to look like green marble, and at night they placed inside lit candles, which shone out from the windows. The ice palace produced "an effect infinitely more beautiful than if it had been built of the most costly marble, its transparency and bluish tint giving it the appearance of a precious stone," one observer wrote.

Shortly, the palace became more than a jewel to gaze at; it was used as the setting for an unusual series of tableaux. A young man and a young maid were dressed in old and colorful Russian peasant costumes taken from a collection of ethnographic displays already accumulated by past tsars, and an ice marriage was mounted to amuse the court. The bridal couple was drenched in water, which formed a light ice coating around them for the duration of the tableaux. After the ice palace had been on display for a while, workmen cut the blocks apart and stored them in the ice cellars of the Imperial Palace, for use in summer months to cool drinks and to refrigerate produce.

In hereditary monarchies such as those of Russia and France, where nearly every sort of commerce capable of yielding large quantities of money was controlled by the crown, the ice trade was a royal prerogative, with monopolies on the gathering and sale of ice granted to court favorites. To properly maintain a courtier at Versailles in the time of Louis XIV required 5 pounds of ice a day. Even in countries where the ice trade was not royally controlled, it was a pleasure reserved for the rich. Only such wealthy Virginia planters as George Washington, Thomas Jefferson, and James Madison could afford icehouses on their farms. In Philadelphia, convicts harvested ice from the frozen Schuylkill River for sale to the public, but few citizens could afford to pay 6 cents a pound. Furthermore, religion-based reluctance to modify temperatures with ice continued; as a contemporary account reports, "Some thought a judgment would befall one who would thus attempt to thwart the designs of Providence by raising flowers under glass in winter, and keeping ice underground to cool the heat of summer."

The reserving of ice for the titled and the rich, the high price of ice, and the religious reluctance to use it found echoes in the scientific laboratories' deliberate distaste for the practical consequences of refrigeration experiments. In 1748 professor of medicine William Cullen did something remarkable in the first chemistry laboratory in Scotland, at the University of Glasgow: he created artificial refrigeration. Basing his work on the recognized phenomenon of cold being produced by evaporating liquids, he tried to intensify the effect by means of a vacuum. Working with "nitrous aether," he exhausted a vessel of its air, which froze the water in another vessel that surrounded the apparatus. Content to write up *Cold Produced by Evaporating Fluids* for publication in a Scottish journal, Cullen did not attempt to exploit the effect commercially.

Neither did English physicist Edward Nairne, who in 1777 improved on Cullen's work by adding another step to the process. As Cullen did, Nairne used a vacuum to evaporate water through a reduction of pressure, but he then employed sulfuric acid to absorb the evaporated water. The absorption process accelerated the rapid drop in temperature of the water in the surrounding vessel and turned that water into ice more quickly than Cullen's simpler process had done. Still, sulfuric acid was so dangerous to work with, and Nairne's process produced so little ice, that while the experiment was frequently replicated in teaching laboratories over the next fifty years, it was not thought of as having commercial applications. It remained little more than a way to demonstrate the "affinity" between sulfuric acid and water. The production of ice was a theatrical grace note that helped students remember and appreciate the general principle.

The next man to take an important step in creating artificial refrigeration had even fewer thoughts of commercialization. Martinus van Marum had wanted to be a botanist, but when blocked from an appointment to that professorial chair at Leiden, he had turned to electricity, then to chemistry, which he studied at the elbow of the acknowledged genius of chemistry in that era, Antoine Lavoisier in Paris. As scientists always endeavor to do, in 1787 van Marum sought to test a hallowed tenet from an earlier era, in this instance, whether Robert Boyle's famous law about the inverse-square relationship between the volume of a gas and the pressure on it held under all circumstances. That was indeed the case if the gas was air, van Marum found; but when he used ammonia, recently isolated by Joseph Priestley, the result was different. After 5 atmospheres — about 70 pounds of pressure — had been applied, each additional twist of the compression apparatus did not produce the expected drop in the volume of the ammonia. Even 7 atmospheres did not decrease the volume any further; rather, it forced the gas to become a liquid.

In liquefying ammonia, van Marum had shown that Boyle's law was not true under all circumstances. He properly concluded that "the aeriform [gaseous] state of whatever fluids ceases to exist, and they are changed into liquids, when they are exposed to the necessary degree of pressure," but he did not test his conclusion's validity by attempting to liquefy gases other than ammonia. Since he was the discoverer of carbon monoxide, he might have tested that, or carbon dioxide, and on the basis of results from three gases replaced Boyle's law with a formula that more adequately described what happens to gases at low temperatures and high pressures.

Of equal consequence to the story of the exploration of the cold, van Marum also did not choose to closely examine the liquid state of ammonia, though it was significantly cooler than the gaseous state. Had he done so, he would soon have recognized that its marked coolness could readily be used to produce refrigeration and ice. Similarly, when several French chemists improved on van Marum's work, liquefying ammonia by means of less pressure and greater cooling of the gas, they, too, did not go on to test whether other gases could be liquefied or to explore the cold that liquefaction produced.

The most powerful factor preventing these lines of inquiry from becoming more fruitful and from blossoming into commercial applications was Lavoisier's theory concerning the nature of heat. The "caloric" theory was a direct descendant of the Descartes-inspired notion of ethereal, eel-like particles that had enraged Robert Boyle. A hundred years after Boyle the concept had been so enlarged that it actually claimed to include Boyle's law on gases, as Newton, in his *Principia*, had suggested that such an elastic-fluid notion could do. In 1787 Lavoisier fully described the elastic fluid he labeled caloric. It was a subtle, weightless, highly elastic fluid believed to flow around or between particles of a gas, holding them in place by means of "repulsive" forces. Caloric was used to explain such disparate processes as the burning of wood in a fire, the heating of water in a metal caldron, the action of the sun's rays, and chemical production of heat. Because of the theory's perceived totality in explaining heat, many scientists — predominantly the French — considered the subject closed and everything that could be known about heat already understood, leaving no reason to further explore it. For the next seventy-five years, researchers encountered considerable difficulty in advancing basic knowledge about heat and cold, until they disproved the caloric theory and substituted other explanations for the source and transmission of heat.

Among the first to challenge caloric directly was Benjamin Thompson, the Count von Rumford. During his service in Prussia, von Rumford had observed that when the solid metal of a cast cannon barrel was bored by a steel drill, high heat was produced in a way that, he contended, had nothing to do with caloric but everything to do with friction. Von Rumford even kept a kettle boiling atop the barrel as it was being bored. The calorists dismissed his attack, using arguments shaped sixty years earlier to refute the friction argument against the existence of an imponderable fluid consisting of "fire particles."

However, in his articles von Rumford expressed in plain words what he understood (somewhat imprecisely) about the principles of heat transmission, and around the year 1800 his articles stimulated the commercial thoughts of a Maryland engineer and member of the American Philosophical Society, Thomas Moore. Moore conceived and built what he called a "refrigerator" and used it to carry butter from his farm 20 miles to the Georgetown market of the then-small national capital city of Washington, D.C. The device was rudimentary — a tight cedar tub, stuffed with rabbit fur as insulation around the edges, with ice within that, surrounding in the center an airtight sheet-metal butter container — but it succeeded in preventing the cool air inside the container from dissipating and thus kept his butter cool. Housewives willingly paid high prices at market for Moore's firm butter, spurning his competitors' products, which, though priced lower, were unrefrigerated and melted to the consistency of grease. Moore patented his device, and in 1803 he published a pamphlet, *An Essay on the Most Eligible Construction of Ice-Houses; Also, a Description of the Newly Invented Machine Called the Refrigerator.* His patent was ineffective, since any carpenter could construct a refrigerator, so Moore made little money from the ice trade. His pamphlet, however, influenced many people, among them the inventor Oliver Evans of Delaware and the commodities trader Frederic Tudor of Massachusetts.

In the fledgling United States of America, any scientific or technological investigation or innovation had to be intensely practical, else no one would pursue it. Oliver Evans had made machines that improved the milling of wool, and steam engines that worked in factories and aboard a dredge; so in 1805, when Evans adapted the work of Moore, Cullen, and Nairne and designed a refrigeration machine that used ether in place of sulfuric acid, and wrote a short book about it, there might have been a reasonable hope that refrigeration would take hold in the United States. But Evans never bothered building his refrigeration machine, deciding it would not bring him as much wealth as would tinkering with steam engines. In that field, too, he lost out to another, better-funded inventor, even though Evans's design for a steamboat preceded Robert Fulton's and was more efficient than his.*

While Evans did not seize the moment for taking the initiative in refrigeration, at almost exactly the same time Frederic Tudor did. The growth of the natural-ice business in the United States is a classic tale, and its protagonist, Tudor, is almost a caricature of the

---

* Evans inspired two other attempts at refrigeration. He corresponded with the English inventor Richard Trevithick, who in 1828 also proposed a refrigerating machine based on Evans's work, a machine that was never built; and he exchanged letters with Jacob Perkins, an American expatriate in London, who patented an Evans-type device in 1834 and for a while made ice from a barge that floated on the Thames. Unable to generate much enthusiasm for his product, Perkins failed commercially; as events would later show, the venture was about thirty years ahead of its time.

iconoclastic, flawed, driven entrepreneur. Frederic was one of four sons of Colonel William Tudor, a lawyer who had clerked for John Adams and served as a judge advocate general during the Revolutionary War, then became a wealthy mainstay of Boston society. The colonel's three other sons attended Harvard, but Frederic considered Harvard to be "a place for loafers, like all colleges," and so at age thirteen, while his parents were abroad, he left the Boston Latin School and began an apprenticeship in a shipping office. At seventeen, he invented a siphon pump for removing water from the holds of vessels and drafted a letter to the Royal Society about it, but never sent it. Short and slight, full of energy, he was an able thinker despite his minimal schooling.

At twenty-one, Frederic accompanied an older brother on a trip through the West Indies, and he returned convinced of the possibilities of making money from that area. After working two more years in the shipping office, trading in pimentos, nutmeg, tea, claret, and other commodities, he was ready to try his own large capitalist venture. At a family picnic in July 1805, another brother, William, half-jokingly suggested to Frederic that the ice that formed on the pond of the family farm at Saugus in winter could be harvested and sold in the Caribbean.

The idea so galvanized Frederic that he bought a leather-bound journal and inscribed on its front a motto that became his credo: "He who gives back at the first repulse and without striking the second blow despairs of success [and] has never been, is not, and never will be a hero in war, love, or business." He dispatched William and a male cousin to Martinique. Both spoke French well — as Frederic did not — and were able to deal with the upper classes to drum up trade for Frederic's later arrival and to prepare a warehouse, based on Moore's designs, to hold the ice. But in March 1806, when Frederic Tudor entered the port of Saint-Pierre on a rented brig carrying 130 tons of ice packed in hay, he found that his advance men had decamped to the Leeward Islands, having made no arrangements for him or his cargo. With the equatorial sun rapidly melting his assets, Tudor first tried to sell ice directly from the boat, distributing handbills to tell purchasers how to preserve and use it. As he wrote to his brother-in-law, the ignorance of people concerning ice was laughable: "One carries it through the street to his house in the sun noon day, puts it in a plate before his door, and then complains that 'il fond' [it melted to the bottom]. Another puts it in a tub of water, a third by way of climax put his in salt!" No one on the island had ever seen or tasted ice cream, and many even had no notion of what an iced drink might feel like; to create sales, Tudor had to first inspire demand. He sought out the proprietor of the island's Tivoli Gardens restaurant and managed to persuade him to sell ice creams Tudor would make. On the first night more than $300 worth was sold, a considerable sum. Tudor wrote home that afterward, the Tivoli Gardens man "became as humble as a mushroom." Still, overall the trip was a failure, resulting in the loss of almost half of his capital investment.

Tudor returned to Boston with ideas for better ice storage in ships and on land, and the next season he managed to sell, at a good profit, a cargo of ice in Havana. "Whenever I find myself run away with by humanity," he wrote to a younger sister, "I feel incompetent to the common duties of existence. *My* son shall be a *fighter*. I will glory in him who shall not be too good. He shall have the common frailties of mankind and as many virtues of the bolder order as shall please heaven to award him." It did not take a mentalist to discern that Tudor was describing himself.

In 1807, blocked from further trade in the Caribbean region because of President Jefferson's embargo on business with colonies of Great Britain and France, Tudor received news of his father's financial ruin. Along with other prominent Bostonians, Colonel Tudor had invested in an attempt to develop land in South Boston; he lost all he had, and thereafter he lived on the small salary he earned as clerk of the Massachusetts Supreme Court. With a sense of the least promising brother becoming the first, and of fulfilling his destiny, Frederic Tudor took on the burden of restoring the family's fortunes.

A martinet who liked to dress in a blue frock coat with brass buttons, Tudor was more than an astute investor; he was also a skilled technologist who succeeded in reducing his warehouse outflow loss from 56 pounds of water per hour to a mere 18. But his prowess could not overcome what he called a "villainous train of events" that included bad luck, and being tricked out of profits by partners in the islands, and imprisonment for failing to pay debts. By the close of the War of 1812, he was near the end of his rope, telling his diary,

> I have manfully maintained as long as I possibly could that "success is virtue." I say so still: but my heart tells me that I don't believe it. Have I not been industrious? Have not many of my calculations been good? [My enemies and bad luck] have worried me. They have cured me of superfluous gaiety. They have made my head grey; but they have not driven me to despair.

"Pursued by sheriffs to the very wharf," as he wrote in his diary, Tudor finally embarked on a successful voyage. He made his first sale, to coffeehouses in Cuba, precisely ten years to the day after the Tivoli Gardens episode, and exulted: "Drink, Spaniards, and be cool, that I, who have suffered so much in the cause, may be able to go home and keep myself warm."

Over the next decade Tudor steadily expanded his enterprise, to Charleston, Savannah, and New Orleans, as well as to more Caribbean islands, creating demand by offering free ice for a month or a season, during which — he proved time and again — the taste for "Northern delicacies" such as ice cream and iced drinks sprouted. Searching for the best packing material with which to separate ice blocks and to keep them dry during transport, he tried hay, rice, straw, twine, and cotton. He also experimented with refrigerating tropical fruits and carrying various return cargoes to New England, but large-scale success still eluded him. He had to bail his brothers out of jail bondage more than once, and in 1822 he experienced what seems to have been a mental breakdown.

In 1823, while Tudor was recuperating in Maine, across the Atlantic one of those accidents occurred that, so frequently in the history of science, presaged an important advance — in this instance, one that would have tremendous scientific ramifications for the next hundred years of the exploration of the cold, and that would also result in commercial refrigeration techniques that would eventually sound the death knell for the natural-ice business. Young Michael Faraday, who had worked his way up from laboratory assistant to stalwart experimenter at the Royal Institution in London, was experimenting at the request of his mentor, Humphry Davy, on exerting pressure on gaseous chlorine hydrate in an attempt to fully explore the properties of chlorine, which had only recently been identified.

Davy did not conduct these experiments himself, among other reasons because they were dangerous. During a single month in 1823, there were three separate explosions in Faraday's lab: one scorched his eyelids, another cut his eyes, the third blew glass into his eyes, which he had to carefully flush out.

On March 6, 1823, a Dr. Paris, Davy's biographer, came in to watch Faraday experiment, and he chided the young man on having a tube containing "oily matter." Faraday agreed that this seemed like sloppy work, and he tried to saw off the end of a pipette to eliminate the contaminated part. In doing so, he must have made a spark that caused a small explosion. When the smoke cleared, the oil in the tube had vanished. The next day, Faraday figured out what had happened and wrote Paris a note about it: "Dear Sir: The oil you noticed yesterday turns out to be liquid chlorine." He also wrote up the results and in a scientific paper claimed credit for the discovery; this claim incensed Davy, who tried to prevent his protégé from being elected to the Royal Society, a blackball that was, fortunately, overridden by the other members. But the wedge between the two men permitted Faraday thereafter to do more work on liquefaction by himself. Shortly, he tackled ammonia.

A real chameleon of a gas, a pungent compound of nitrogen and hydrogen, at one moment ammonia could exist in the liquid compound "ammonia water," and at the next — when the ammonia water was heated — ammonia would separate from the water and become a gas; and that gas could also be subjected to pressure and turned into a different liquid, a liquefied gas. In 1823 Faraday examined all of these states, and in the process, he demonstrated that liquefied ammonia could be used to generate cold.

To understand why, and how, it is necessary to repeat, in print, an ice experiment that many scientists were performing for themselves and their students in laboratories all over Europe in the first quarter of the nineteenth century. Its point was to show how ice cools things — not, as most people believed, by conduction alone, by conveying its own coldness to neighboring materials — but mostly by melting, during which ice absorbs heat from its surroundings. Scientists first mixed a pound of water at the boiling point, 212°F, with a pound of water at 34°F; the result was 2 pounds of water at 123°F, a midpoint that could be mathematically expected. But then they mixed a pound of water at 212°F with a pound of *ice* at 32°F and obtained a startlingly different result — 2 pounds of water at 51°F. An amount of heat that would have been sufficient to raise 2 pounds of water another 72°F had been absorbed by the change from solid ice to liquid water. Faraday recognized that what made ammonia change rapidly from its liquefied to its gaseous state was a similar absorption of heat from the surroundings — and that this was what lowered temperature and produced cold.

Faraday noted that "there is great reason to believe that [this technique] may be successfully employed for the preservation of animal and vegetable substances for the purposes of food," but he did nothing personally to advance the commercial possibilities of generating cold by means of ammonia's absorptive capacities. The reasons for this inaction are not described in his notebooks or diaries, but he was known to be a determinedly uncommercial man, a member of a Quaker-like religious sect devoted to giving up worldly things to better elucidate the works of God. During the first flush of the Industrial Revolution, Faraday turned down repeated requests from entrepreneurs to act as a paid consultant to their endeavors in other fields, and surely he would not have considered setting up a refrigeration business or licensing his process to others to do so.

Simultaneously with Faraday's experiments, in 1823 in Paris another, more obscure experimenter was working. An École Polytechnique graduate, Charles Cagniard de la Tour, was attempting to do the opposite of what Faraday was doing: turn a liquid into a gas. By using heat and pressure, Cagniard de la Tour succeeded in pushing pure alcohol to the point where it completely became a gas. By 1832 he had done enough work to reach the important conclusion that each liquid has a "critical temperature" above which it must pass into a gaseous state. This conclusion also meant the obverse was true: a temperature existed below which every gas must become a liquid.

The chemist Charles Saint-Ange Thilorier followed up on the work of Cagniard de la Tour. In the laboratory of the School of Pharmacy, in Paris, in 1834, Thilorier accomplished what other researchers had tried but failed to do: he pressured gaseous carbon dioxide until it became a solid, carbonic acid, which afterward was known as "dry ice." The accomplishment did not come cheaply: one experiment resulted in an explosion in which Thilorier's assistant lost both legs. Dry ice would eventually come to be a major component of refrigeration systems. But though Thilorier was working in a school of pharmacy, he evidently thought no more about commercial possibilities than Faraday had. What he did do with solid carbonic acid was stir it into a mixture of snow and ether, and thereby reach a temperature of $-110°C$, the lowest ever recorded at that time.

By a single technological leap, a far interior outpost of the country of the cold had been gained. Minus 110°C was about as far below the freezing point of water as the boiling point of water, plus 100°C, was above it. Moreover, −110°C was a temperature that had not previously been proved to exist, though scientists had understood from Amontons that such a low temperature was possible, and it existed nowhere else on Earth except in this Parisian laboratory. Despite that, the Thilorier work did not excite much attention among other scientists, and it languished for lack of interest, as did other scientific work equally important to the history of cold, also done in Paris in that era, by the theoretician Sadi Carnot, as will be recounted in the next chapter.

Back in the American Northeast, in the mid-1820s, Frederic Tudor recovered from his mental breakdown and made two important alterations to his personal and business life. He married, and he had the good sense to hire young Nathaniel Jarvis Wyeth. The two men shared an aversion to college and a predilection for intellectual companions — editors and publishers. Wyeth was not a theoretical scientist, or a scientific experimentalist, or an instrument maker/technologist whose work was based on advances in scientific understandings. He was a pure tinkerer whose 1825 device for harvesting ice revolutionized the industry. A saw-toothed cutter designed for a horse to pull, it scored the ice so deeply and neatly that the ice could then be readily split into blocks and floated downstream to an icehouse. The ice cutter slashed the cost of harvesting ice from 30 cents a ton to 10 cents; also, because it produced extremely regular blocks, it made shippers more eager to carry these blocks than they had been to transport the irregularly shaped blocks and chips of old, which could shift about dangerously in a hold. Tudor first contracted with Wyeth to supply ice, then, recognizing Wyeth's worth, employed him directly. Tudor was soon confiding to his Ice-House Diary that on a visit to his ice-harvesting site, he had been pleased to chance upon his new employee "wandering about the woods at Fresh Pond in all the lonely perturbations of invention and contrivance. His mind evidently occupied in improving the several contrivances which he is perfecting for carrying into good effect improvements in his several machines for the ice business."

Among those advances was the determination of the best insulation between blocks of ice. Wyeth fastened on a ubiquitous substance whose use for this purpose had eluded everyone else: sawdust, from the hundreds of sawmills throughout New England. With this innovation, and Wyeth's series of technological advances — an endless chain to lift blocks from the river channel, an auger to drill through fields of ice and drain water on top, new tongs, new tools — Tudor's enterprise shot ahead rapidly, and by the early 1830s he almost monopolized the ice trade in the United States.

Tudor doubled Wyeth's salary, to $1,200 a year, but did no more for the younger man, even as the business expanded so much that Tudor became known as the Ice King. Wyeth wanted to go out on his own but lacked the capital to start an ice-trade enterprise. The two men struck a deal: Tudor paid Wyeth $2,500, and in exchange obtained Wyeth's patent on his ice-harvesting process. Wyeth then left for Oregon, expecting to procure furs and salmon to sell in the East. His trip ended in frustration, but his work was so audacious and his reports of it in the journals he kept were so vivid that Washington Irving eventually turned them into a popular book, *The Adventures of Captain Bonneville.* Tudor discovered that having the patent without the inventor did no good, since rival ice producers used Wyeth's methods without payment to Tudor, and since in the absence of an inventive second-in-command, Tudor was unable to prevent the rivals from cutting into his business in the South. Tudor also exacerbated his losses through an ill-conceived venture in coffee trading, in which more than $150,000 vanished.

Broke again, Tudor escaped jail after promising to continue in business and repay every cent he owed, plus interest. Wyeth came back from Oregon and toiled again for Tudor, but then the two men locked horns in an acrimonious court dispute over the rights that had been conveyed in the patent sale, and they became rivals. Wyeth helped charter a railroad line from the ice ponds to the Boston

docks, and he introduced steam power to the harvesting, warehous-
ing, and transport of ice, reaping from these efforts a good but not
enormous income. At the same time, Tudor's emissaries successfully
ventured to India, other parts of southern Asia, Australia, and South
America. Henry Thoreau watched Tudor's cutters working on Walden Pond and marveled that water from his bathing beach was
traveling halfway around the globe to become the beverage of East
Indian philosophers. In 1849 a Scottish journal, citing the successful
delivery of ice to Calcutta, suggested, "If means could be contrived
for transporting fresh meat in ice at small cost, Europe would present a steady market for the surplus beef and mutton of America."
By the outset of the Civil War, Tudor, Wyeth, and their competitors
were shipping the equivalent of one cargo each day, every day of
the year, to more than fifty ports. Tudor repaid what he owed and
became a multimillionaire, principally responsible for an annual ice
harvest in the Northeast measured in the hundreds of thousands
of tons.

Richard O. Cummings points out in the definitive study of the
American ice trade that Tudor made more money in the latter part
of his career under conditions of competition than he did when he
had a virtual monopoly on natural ice. The reason for Tudor's rise in
wealth was the vastly enlarged dollar volume of the ice trade itself.

Most of the ice used in the world was consumed in the United
States. Starting in the 1820s, ice consumption more than doubled
each year as Americans grew used to having ice to cool drinks and to
keep food fresh in larders. When various innovations permitted
Tudor, Wyeth, and their rivals to reduce the retail price to 12½ cents
for 100 pounds — a far cry from the $6 per 100 pounds that ice had
once cost — its cheapness spurred further innovation in food use,
which in turn led to even greater demand for ice. One of the new
concoctions was to become a totem of the South, the mint julep, and
it is entirely likely that the first of these alcohol-laden drinks, in the
1820s, was made with ice from that totemic site of the Northeast,

Walden Pond. Shortly, many southerners sought ice for making juleps. Temperance societies also liked and touted ice, believing it promoted the use of a more healthful beverage, cold water. Ice consumption in New Orleans rose from 375 tons in 1827 to 24,000 tons in 1860; and during the same period, New York City raised its annual consumption of ice to 100,000 tons.

The diets of urban Americans underwent a marked change starting in the 1830s, a change intertwined with the increased use of ice. Fresh, unpreserved fruits, vegetables, meats, and milk rose in popularity. Upstate New York farmers began using ice to refrigerate their milk, so it could be taken by railroad to New York City, a trip of more than four and a half hours. Fresh seafood was iced and transported from port cities to cities 200 to 300 miles inland. Ice also prolonged the period of the year when meat could be cured, providing the steady low temperatures necessary to preserve meat with salt or smoke before it began to spoil.

Rapidly enlarging American cities also contributed to the growth of refrigeration in two further ways. An expanding urban population that could not grow its own food needed refrigeration to preserve its supplies; and with fewer people left in the rural areas to harvest the fields and transport the products quickly, ever more refrigeration was needed to store food until ready for market.

Most refrigeration occurred in the country's North and Midwest. The relative dearth of it in the South, because of the absence of natural-ice sources and the expense of importing ice, affected the character of farming in the region. While the northern areas developed dairy, vegetable, and meat farms, the southern farms, with less access to and facilities for refrigeration, tended to stick with cotton and tobacco, crops that could be grown, harvested, stored, and transported to market without temperature controls.

Refrigerators that used stored ice were only partly effective, because refrigerator makers did not yet understand the way that ice cools. They did not comprehend that ice does not cool by conduction but by absorbing heat from the surroundings. Lacking that knowledge, early refrigerator makers like Moore constructed their devices to preserve ice and preclude it from melting by cutting off the airflow, an action that prevented the ice from cooling by absorption. Not until 1845 did inventors begin to produce icebox refrigerators that circulated air and therefore worked better than the older ones.

Another spur to American demand for refrigeration was the introduction of lager beer. Before lager arrived from Germany, American beers were made by "top-fermenting," a process that took place at the surface of the liquid and could occur at any temperature. Lager beer was created through the action of yeast, in a bottom-fermenting process that resulted in a mix that then had to be stored at between 47° and 55°F so it could mellow, develop carbon dioxide fizz, and become an aged beer. Ice could ensure the maintenance of the desired temperature and also permit the manufacture of lager beer year-round, instead of only in winter. The arrival of significant numbers of German immigrants to the United States in the 1840s accelerated the demand for lager beer and for ice to produce it. By the 1860s, American brewers were buying $1 million worth of ice each year.

"Ice is an American institution — the use of it an American luxury — the abuse of it an American failing," *De Bow's Review* asserted in 1855, contrasting the American use of ice for household refrigeration with that of Europe, where ice was "confined to the wine cellars of the rich, and the cooling pantries of first class confectionaries." Ice helped point the American dream in the direction of material comfort for the masses. And ice became a symbol of America: when a competitor of Tudor's broke into the London market, he mounted a specimen of his ice, taken from Lake Wenham in Massachusetts, in the window of his London store, replacing it each day so that to passersby the ice appeared crystalline perfect, never melting, always the same. Shortly, Wenham ice became synonymous with purity and was greatly desirable — the empress Anna's jewel of an ice palace, made accessible to the masses by the power of American democracy and unfettered capitalism.

> We know of no want of mankind more urgent than a cheap means of producing an abundance of artificial cold. To warm countries it would afford benefits as countless in number as those that arise in cold climates from the finding of illimitable supplies of fuel. The discovery and invention which our correspondent proposes to apply to this object are calculated, if true, to alter and extend the face of civilization, and we trust that a measure which promises to be attended with such results will not be suffered to be neglected, or fall into oblivion.

Thus wrote the editor of the *Commercial Advertiser*, in an 1844 editorial response to a series of eleven articles by someone named Jenner. No one would have proffered that opinion earlier in the century, before the use of natural ice had spread to the middle class; the growing use of natural ice stirred into being the wish for a way to produce ice whenever and wherever a need or a yen for it arose. In this instance, invention — in the form of facilitating the natural-ice industry — was surely the mother of the new necessity to manufacture ice to meet the heightened, "urgent" demand.

The Jenner name was a pseudonym. The articles in the *Commercial Advertiser*, and the prospective ice-producing machine, were the products of Dr. John Gorrie, the leading physician of Apalachicola, Florida. Born in Charleston in 1803 of either Spanish or Scotch-Irish extraction, he studied at a medical college in western New York, and after an internship elsewhere, he settled in Apalachicola in 1833. He quickly became the port's leading physician, its postmaster, a member of its governing council, and, in 1837, its "intendant," or mayor. After two years as intendant, he retired from public office to devote himself exclusively to medicine and science. Evidently at the urging of Senator John C. Calhoun, who was concerned about a naval hospital in Apalachicola that cared for sailors ill with malaria and

yellow fever, Gorrie accepted a contract to supervise this hospital.*
In the early 1840s he conceived a project to cool the hospital's air,
believing this would help cure feverish patients and possibly even
prevent malarial diseases from spreading. He planned to artificially
produce ice to cool the hospital, by a process he would describe in
his patent application as drawing on the "well-known law of nature," that compressing air produces heat and expanding air produces cold, the latter effect being "particularly marked when [air] is
liberated from compression."

What Gorrie referred to as a well-known law of nature was hardly
understood beyond a handful of scientists, and no one else had
picked up on its commercial potential, or attempted to make ice on
the grand scale necessary to cool entire rooms. The notion so intrigued Gorrie that by 1844 he entirely abandoned his medical practice to pursue it. Yet his ideas were considered so outlandish and
heretical (in that they might contravene God's plan of the world's
hot and cold regions) that Gorrie felt impelled to write those eleven
articles under the alias.

The editor of the *Commercial Advertiser* praised the unknown
author in his editorial, but he also lightly chastised him for not yet
fulfilling the "moral obligation" to go beyond theory and make a
useful device. What the editor seemed not to know was that Gorrie
was already using a device to cool two special hospital rooms and his
own home. His first device suspended a basin with ice from the
ceiling of a room and blew over it fresh air carried down the chim-
ney by a pipe. In 1849, after Gorrie had worked for five more years to
perfect a working model of an ice-making machine, he first applied
for American and British patents.

The summer of 1850 arrived early to New England, prematurely
melting the ice on the ponds and rivers, so there was less ice to ship
south. Apalachicola was without ice, an abominable inconvenience

---

* Gorrie also recommended draining the swamps to remove the local cause of malaria,
even before the principal transmitter of the disease, the swamp-born mosquito, had been
identified as its carrier.

for the guests of the Mansion House, then the largest hotel in Florida. When one cotton buyer wished for ice for his wine at dinner, another buyer, a Monsieur Rosan of Paris, bet him a bucket of champagne that not only could he furnish the ice, he could make it right in the dining room. Rosan had been working with Gorrie, who took the occasion of this wager to make the first public demonstration of the machine. News of Gorrie's accomplishment reached New York City, where the *Globe* commented, "There is a Dr. Gorrie, a crank down in Apalachicola, Florida, that thinks he can make ice by his machine as good as God Almighty."

Meanwhile, the British granted Gorrie a patent in 1850, which was reported in a laudatory article about his process in a British publication — spurring German technologist William Siemens to order one of the machines and to design a similar but slightly improved process. Gorrie received his American patent in 1851, but none of these events resulted in his attracting the financial backing necessary to manufacture a large machine that could produce commercial quantities of ice. He therefore went to New Orleans to find backers. Bankers there refused him, citing the ready availability of natural ice transported by ship from the Northeast. Next, he sold a half interest in his invention to a Boston investor, in exchange for expected cash; but the man died shortly after signing the deal, and Gorrie returned home to Apalachicola without the money. Finally, after publishing *Dr. John Gorrie's Apparatus for the Artificial Production of Ice in Tropical Climates* in 1854, he contracted an illness and died the next year, with his commercial machine not yet produced.

Gorrie had an American rival for primacy in the artificial production of ice. Alexander Catlin Twining, the son of a Yale official, studied for the ministry before becoming enamored of mathematics and switching to West Point, where he studied civil engineering and astronomy. Observing a spectacular meteor shower in 1833, he formulated a theory of the cosmic origin of meteors that countervailed the then-current assumption that meteors lived and died within the earth's atmosphere. After a stint of railroad engineering, Twining accepted the chair of mathematics and natural philosophy at Middlebury College in Vermont. He became interested in producing ice after doing some experiments in the 1840s, with a process that centered on condensing ether vapor, and by 1849 he had resigned his chair to develop a commercial ice-producing machine. With investors, he had a plant constructed in Cleveland, Ohio, in 1853; he detailed his process in his 1857 booklet, *The Manufacture of Ice on a Commercial Scale*. Had Twining established his plant in, say, Atlanta, he might have enjoyed commercial success, but by placing it in a northern city that had access to ice from the Great Lakes, he virtually ensured that natural-ice marketers would purposely lower their prices to prevent his ice from replacing theirs. Like Gorrie, Twining died a bitter man, his ice-producing dreams never fully realized.

Gorrie and Twining were shortly eclipsed by the French entrepreneur Ferdinand Carré. Twenty years after Faraday's ammonia-absorption experiments, Carré adapted them to make an "absorption" refrigeration machine. It began the procedure by applying heat and pressure to *aqua ammonia*, which separated the ammonia gas from the water. The gas then passed into a condenser of pipes containing cold water; in this cooler environment, a second application of pressure liquefied the ammonia gas. The liquefied gas then flowed into the actual refrigerating chamber, also known as the evaporating chamber because this was where the liquid ammonia was forced to evaporate — to become a gas again, and to expand as it did so, absorbing heat and producing the "cold effect" that turned water in an adjoining chamber into ice. After the gas had done all of this, it was reabsorbed into the first batch of water, becoming, once again, *aqua ammonia*.

A Marseilles brewery installed Carré's prototype machine in 1859, and in 1860 he won patents in France and in the United States. Then came his big break — the onset of the Civil War in the United States. Because the Gorrie process had languished after its inventor died, and because Twining was a Northerner whose machinery was not welcome in the South, the Civil War provided an opportunity for Carré. Several of his machines were shipped past Union blockades into Southern ports and set up to produce ice, where the Northeast natural-ice traders were no longer supplying ice. Southerners used Carré-process ice principally in hospitals but occasionally to provide ice-cooled delicacies that permitted some households to maintain the illusion that the war had not affected their lifestyles.

The wartime success of the Carré plants proved the efficacy of artificial icemaking and set the stage for the spread of ice-based refrigeration throughout the world. But to achieve further mastery of the cold would require more precise understanding of its basic processes — and the search for them had been under way, by a group of unlikely scientists, for some time.

The Confraternity
of the Overlooked

THE EARLY NINETEENTH CENTURY WAS a hectic and confused time in science, and research into the nature and uses of cold suffered from science's inadequacies. Some of the confusion stemmed from societies having to deal with the upheavals of the French and American revolutions; scientists also had to struggle against a deeply entrenched, mechanistic conception of the world that had solidified 150 years earlier. It had been articulated by Robert Boyle and his generation, for instance in Boyle's characterization of the universe as "nothing but . . . a machine whose workings are in principle understandable by human reason . . . [like] a rare clock . . . where all things are so skillfully contrived, that the engine being once set a-moving, all things proceed according to the Artificer's first design." In that clockwork universe, light was considered an invisible substance composed of corpuscles, and chemicals were attracted to or repulsed by one another because of natural affinities and molecular forces, one of them being the "subtle fluid" called caloric, believed responsible for heat and cold by combining with other substances in unfathomable ways. These mistaken notions had to be overcome before scientists could make progress in basic understandings of how the universe actually works.

But in the years around 1800, science had not yet entirely disentangled itself from either magic or philosophy. Audiences filled popular lecture halls to see and hear chemists, partly because they provided spectacular demonstrations and explosions, and partly in the hope that chemistry would confirm or refute the philosophic doctrine of materialism, which insisted that man had no immortal soul, and matter was just matter. Coleridge attended such lectures. Goethe wrote a novel using a chemistry-based metaphor, *Elective Affinities.*

Physicists, who considered chemists no more than apothecaries, disdained the study of heat and cold as belonging to chemistry, since heat was believed to be a product of chemical reactions connected to oxygen burning. Chemists believed that oxygen burning and the theory of caloric had explained everything necessary to know about heat. With both physicists and chemists unwilling to investigate the phenomena further, heat and cold became the least desirable field of inquiry for scientists just at the very moment when heat, in the form of steam engines, was revolutionizing the labors of humanity.

It was the genius of an engineer, Nicolas Léonard Sadi Carnot, to unite the study of steam engines and the study of the fundamental physics of heat and, in the process, to lead the way to understanding what cold is and how it is produced. His single published work, *Réflexions sur la puissance motrice du feu (Reflections on the Motive Power of Fire)*, a study of an ideal steam engine published in 1824, would eventually be praised as among the most original works ever written in the physical sciences, with a core of abstraction comparable to the best of Galileo. It would greatly influence the study of what came to be called thermodynamics, and in the twentieth century it would form the basis for constructing apparatus to reach within a few billionths of a degree of absolute zero. But during Carnot's lifetime, the book was virtually ignored.

In the 1870s, when Hippolyte Carnot found some old notes of his long-dead brother and convinced the Académie des Sciences to print them, he began his accompanying biographical sketch of his brother by writing that "the existence of Sadi Carnot was not marked by any notable events." But he also provided tantalizing descriptions of a man of "extreme sensibility, extreme energy . . . sometimes reserved, sometimes savage," who studied such diverse things as boxing, music, crime, and botany. Among the maxims that Hippolyte extracted from Sadi's notebooks was this mordant gem: "It is surely sometimes necessary to abandon your reason; but how do you go about retrieving it when you have need of it?"

One school of scientific biography contends that a scientist's accomplishments can only be understood by reference to his times; another school, by reference to his personality. In the case of Sadi Carnot, both his era and his personality were deeply influenced by his father. A mathematician, engineer, and soldier, Lazare Carnot published in 1783 an essay that discussed the dynamics of machines in terms of the "work" they did, rather than in terms of forces, à la Newton. Crossing into politics, by 1793 Lazare had risen to share with Robespierre and others of the Committee for Public Safety the responsibility of establishing fourteen armies for the revolution, as well as for condemning people to the guillotine. Exiled for opposing a coup, he came back to power with Napoleon in 1799, lost his positions in 1807, was recalled to service in 1814, and after Waterloo was once again exiled.

A founder of the École Polytechnique, Lazare helped arrange for the acceptance there of his eldest son, Sadi, as a pupil-cadet in 1812, at the age of sixteen. Sadi won a first prize in artillery, and after taking part with classmates in the siege of Paris, he transferred to the École du Génie, the school for artillery and engineering, where he wrote a paper on an astronomical instrument and toiled on fortifications through the remainder of the last Napoleonic War. Denied promotion in the army, he was seldom employed in the specialty for which he had trained. "Fatigued with the life of the garrison," as Hippolyte put it, Sadi transferred to the general staff in Paris in 1818, then retired at half pay into "voluntary obscurity."

He gravitated to the Conservatoire des Arts et Métiers, a recently founded technological museum that scandalized establishment scientists by offering lectures to the general public. In addition to being a temple of the practical, the Conservatoire was a hotbed of liberal, antiroyalist sentiments, which Sadi shared. Among its stalwarts were two men who came from the same region as Carnot, Nicolas Clément and Charles Bernard Desormes, brothers-in-law who did joint research on the physics of steam engines. All three felt keenly France's military loss to England, understood that French industry needed a boost to prevent England's burgeoning mills and factories from eclipsing France's own, and believed that the way to improve industry was to better understand the principles behind the operation of machines.

Sadi Carnot devoted the years from 1820 to 1824 to the 118-page *Réflexions*, which he self-published in an edition of 600 copies. The book remained obscure until well after his death. Among the reasons for its being ignored during his lifetime: it was not written by an Académie-certified expert, was not published in an establishment periodical, and did not contain original experimental results. Pointing out that the steam engine had become more important for the economic health of England than its navy, Carnot set out to explain why the newer steam engines were more efficient than James Watt's original, to postulate the maximum efficiency of an engine under ideal conditions, and to deduce from that inquiry the general relationship between heat and mechanical work. His thesis was that the action in the steam engine was a function of temperatures and that the power of the engine had to do with the fall in temperature from hotter to colder. He drew on earlier attempts — by his father, among others — to improve the water-wheel engine, in which power derived from the volume of water and the length of time it was in contact with the wheel, during which the water was carried from high point (entry) to low point (exit). The action in the steam engine, he insisted, was analogous, the "motive power" of its heat depending "on what we shall call the height of its fall, that is, on the temperature difference of the bodies between which the caloric flows."

Carnot's central tenet was that mechanical work was produced in proportion to the fall (of caloric) between higher- and lower-temperature bodies. He undermined his argument somewhat by relying on the theory of caloric. Antoine Lavoisier, the father of French chemistry, had died on the guillotine in 1794, but his theory of caloric had lived on. Carnot was actually uneasy with caloric; acknowledging critics of caloric such as Count von Rumford and Humphry Davy, he wrote that the basic principles of the caloric theory of heat needed "close attention" because "many experimental results would seem to be nearly inexplicable according to the present state of the theory." But he also balanced their criticism by citing a series of prizewinning French experiments done in 1812 on the specific heat of gases in relation to their density, the results of which seemed to shore up the notion of caloric.

Unknown to Carnot, the figures obtained in the 1812 experiments were wrong, an error that the historian of caloric, Robert Fox, calls "one of the most influential in the whole history of the study of heat. Backed by the prestige associated with victory in [an official] competition, the result quickly became standard and . . . misled many calorists."

With hindsight, we can see that Carnot's basic discovery did not depend on caloric: he asserted that mechanical work could not exist unless heat was transferred from a body at a high temperature to a body at a lower temperature, and that the greater the temperature difference between those two bodies, the more work was done. Carnot's four-cycle ideal engine produced the same work backward as forward; this alternation was crucial to his contention that the maximum amount of work possible was done in an engine whose processes were reversible. But while the sequence of strokes might be reversible, the direction of the flow most emphatically was not. Donald Cardwell, a modern historian of thermodynamics, points out that only Carnot, of all those who wrote about engines in this era, had the genius to recognize that "the vast majority of thermal and thermo-mechanical changes are . . . irreversible."

A quarter century later, after the theory of caloric had been disproved, that irreversibility would lead Rudolf Clausius and Lord Kelvin (William Thomson) to formulate the second law of thermodynamics; moreover, in the late twentieth century, Carnot's understandings of the working of the ideal steam engine would lead to many advances in the generation of ever lower temperatures achieved by means of a "Carnot cycle" used to produce cold close to absolute zero.

Among the reasons Carnot could not accept all the implications of the irreversibility he postulated was that he agreed with a pillar of the mechanistic universe, the notion of the conservation of all matter and forces. This idea contended that everything in the universe was already in existence, and nothing could be created or destroyed. It was a belief with religious origins, and Carnot could not afford emotionally to accept the damage to it that his own insight would bring. The idea that matter could indeed be irrevocably destroyed or changed in some not-yet-understood way was a frightening concept to this otherwise clear-eyed scientist.

Carnot presented *Réflexions* before the Académie. There was a long and appreciative review of it in one journal, a brief notice in another, and an encomium by Clément recommending it to students, but the book brought the author no renown. After the publication, Carnot was briefly activated again in the army, then returned to Paris, where he described himself in 1828 as a "constructor of steam engines." His studies, now more specifically dealing with the physics of gases, were interrupted by the revolution of 1830, which toppled Charles X, restored a degree of popular sovereignty, and established a new king. Carnot became part of a cohort of École Polytechnique graduates who supported the new order. In the spring of 1832 an inflammation of the gorge confined him to his bed; by summer he was so weak that he could not fight off cholera, and he died in August. That Carnot died at Charenton, a hospital associated with the insane, occasioned some historians to say he went mad; but Charenton was used for cholera patients in 1832 because other hospitals were overcrowded, and in the hope that isolating those with cholera would halt the progress of the disease through the population.

Unknown at the time of Carnot's death, but of importance to our story, was that around 1830 he had come to the realization that the caloric theory was wrong. The corpuscular theory of the transmission of light had been disproved, and experiments were demonstrating the likelihood that electricity, light, and magnetism were not the products of separate "forces" but were interrelated. In terms of heat and caloric, Carnot asked, "How can one conceive of the forces agitating the molecules, if they are never in contact with one another, if each [molecule] is perfectly isolated? Supposing that there is a subtle fluid interposed doesn't reconcile the difficulties, because that fluid would necessarily be composed of molecules." This reasoning led to an important conclusion:

> Heat is nothing other than motive power, or perhaps motive power that has had a change of form. If there is a destruction in the particles of a body, there is at the same time heat production in a quantity precisely proportional to the quantity of motive power that is destroyed; reciprocally, in every configuration, if there is destruction of heat, there is production of motive power.

This understanding of how heat was transformed into motive power pushed Carnot to boldly state a modification of the everything-in-nature-is-preserved notion that he had been unwilling to make in 1824 but that he could no longer avoid:

> that the quantity of motive power in nature is invariable, that it is never properly speaking produced nor destroyed. Truly it changes form, sometimes manifesting itself as one kind of movement, sometimes as another, but it is never annihilated.

Carnot was reaching here for a concept that he could not elucidate and that would take another quarter century to be defined and understood: energy. It is energy that is never annihilated, merely manifested as one or another kind of movement. Carnot didn't quite get the concept right, but he came close. In 1830 it would have been shocking to contend publicly that heat was not conserved, because it would call into question all of French science based on the work of Lavoisier and Marquis Pierre Simon de Laplace, considered France's greatest mathematician. Equally disturbing would have been Carnot's new contention that matter could be completely transformed into motive force, which countermanded the idea that the Artificer of the universe would permit some aspect of His creation to vanish into thin air. Cardwell suggests that Carnot's most compelling reason for not publishing these notes during his lifetime, though, was that to do so he would have had to revise his major work in the light of his new understandings, and that task was beyond the capacity of a single individual; the complete revision and integration of his new ideas would required the combined talents of some of the century's most ingenious thinkers.*

The French engineer Émile Clapeyron was a contemporary of Sadi Carnot, having passed through the École Polytechnique a few years after him, and having possibly been in touch with him during the 1830 uprising. Two years after Carnot's death, Clapeyron published an exegesis of Carnot that did what Carnot had been at pains not to do: it used mathematical formulas, graphs, and diagrams, for instance, to extrapolate from Carnot's analysis "Clapeyron's equation," which holds that the maximum amount of work a unit of heat can perform when it vaporizes a liquid cannot be larger than the amount it would perform if it were doing another task. Clapeyron also proved mathematically Carnot's contention that the amount of work done during the course of a one-degree fall of a unit of heat

* Carnot's rejection of caloric in his post-1824 notes was one reason why the Académie des Sciences was eager to publish them in 1878, even though it had ignored Carnot during his lifetime. The second reason was that the notes showed clearly that a Frenchman had preceded all the German and English scientists in coming to understandings of both the first and second laws of thermodynamics.

decreases as the temperature increases. More widely circulated than Carnot's book, Clapeyron's paper was also translated into English and German, which made it accessible to those doing research in what would become known as thermodynamics, the study of the transformations of energy, leading to its further influence on the exploration of the cold.

As pointed out in the previous chapter, during the 1820s and 1830s science's growing ability to produce cold was ignored. In the same year as Clapeyron's publication, 1834, and also in France, an amateur scientist of independent means used electricity to directly produce heat and cold in a way that in the twentieth century would become quite important. Jean-Charles-Athanase Peltier, who had retired from clockmaking when the death of his wife's mother resulted in a small inheritance that allowed him to follow his scientific interests, had been intrigued by the research of an earlier Estonian-born German physicist, Thomas Johann Seebeck. Peltier passed a continuous current along a circuit of two conductors, made from different metals and connected by two junctions, and found that the temperature of one junction rose and the temperature of the other junction fell.* In other words, electricity could be used either to cool or to heat: Seebeck and Peltier had discovered an entirely new field, thermoelectricity. In 1838, using Peltier's thermoelectric method, the German scientist H. F. E. Lenz froze a drop of water. He, too, was ignored. In other words, by 1838 the technical means of providing cold to those who might need or want it had been demonstrated in several scientific laboratories, but neither the pure scientists, nor the technologists, nor the few would-be entrepreneurs of refrigeration seemed interested in the process or the goal.

Aboard the Dutch ship *Java* in Jakarta in 1840, ship's doctor Julius Robert Mayer observed something unusual while trying to prevent the spread of an infection through his crew. As he later recalled,

---

* The explanation for this eluded the generation of Seebeck and Peltier and was approached later by Kelvin. See chapter 6, page 104.

"The blood let from the vein in the arm had an uncommon redness, so that from the color I could believe I had struck an artery." Mayer, the twenty-six-year-old son of a German apothecary, had an unusual mind, by turns highly religious and full of humor, and was given to doing card tricks, solving rebuses, winning at billiards, and writing aphorisms — "What is insanity? The reason of an individual. What is reason? The insanity of man." Mayer decided that the blood of his shipmates was brighter red in Jakarta because in the hotter climes, human bodies required a lower rate of oxidation, and since their bodies extracted less oxygen from food, their venous blood was redder; this led him to ponder the relationship between the body heat of an animal and the work done by that animal — which, in turn, spurred him to think about the relationship between heat and work in any configuration of matter, organic or inorganic.

Back home in Heilbronn, in February 1841 Mayer collected his thoughts in an article that he sent off to the leading German natural-science journal. It included this typical, nearly unfathomable sentence:

> If two bodies find themselves in a given difference [chemical difference or spatial separation], then they could remain in a state of rest after the annihilation of [that] difference if the forces that were communicated to them as a result of the leveling of the difference could cease to exist; but if they are assumed to be indestructible, then the still-persisting forces, as causes of changes in relationship, will again re-establish the originally present difference.

This actually meant something important — it was a statement of what would later be called the first law of thermodynamics, that energy cannot be destroyed, it can only be converted to other forms — but it was almost impossible to understand from Mayer's writing. Mayer also harmed his cause by couching his argument in the context of Immanuel Kant's phenomenological approach to the study of nature, suggesting that indestructible forces such as heat and mechanical work were halfway between inert matter and soul. This was precisely the sort of philosophic blather that serious physicists were endeavoring to move beyond, and so Mayer's article was ignored by the journal, as were his three follow-up letters to the editor about it.

His next article was somewhat clearer. In it, he tried to "answer the question of what we are to understand by 'forces,' and of how such [forces] are related to each other." This provincial doctor, untrained in physics, proceeded to cast doubt on the foundations of the clockwork universe by contending that Newton had misperceived gravity. Gravity was not a force but a property of matter, Mayer said, and Newton had confused the two concepts. The qualities of a force were "indestructibility and transformability," and gravity had neither of these; nor could gravity cause motion, because movement also required spatial separation of objects. That spatial separation of bodies, which Mayer labeled "fallforce," was the real force. Motion, heat, and fallforce, Mayer wrote, were different configurations of the same "indestructible, transformable, imponderable" substance.* He tried to come up with a numerical measurement of that substance by analyzing others' experiments, and deduced a figure: "the warming of a given weight of water from 0° to 1°C corresponds to the fall of an equal weight from the height of about 365 meters [1,200 feet]." This seemed to most readers like a comparison of apples and oranges.

This article was published, but it was ignored by serious students. Years later, Rudolf Clausius would confess to Mayer that he had not bothered to read it because its title did not include the word *heat* or *motion*. When Mayer produced another article in 1845, he had to pay for its printing himself. In it he equated heat with mechanical effect and contended that both were independent of whatever fluid or gas was being used, since the fluid or gas "serves only as a tool for effecting the transformation of the one force into the other." C. A. Truesdell, III, another modern historian of thermodynamics, has

---

* By "imponderable," Mayer meant weightless.

called this a major theoretical insight, the interconvertibility of heat and work. Carnot had crept up on it but had not stated it very well; Mayer did it better. Within a decade, this idea — expressed by others — would lead to advances in producing seriously cold temperatures. Truesdell adds that because Mayer was "virginally innocent" of the theories and experiments of mainstream scientists, he did not properly frame his interconvertibility insight in mathematical terms, which resulted in its remaining unknown and unheralded.

Steam engines, boilers, locomotives, and factories based on steam power were the elements of the workaday world of Manchester, England, and Glasgow, Scotland, in the first half of the nineteenth century, when these cities were at the center of the Industrial Revolution. The leading scientists born and raised in those cities in those years — James Prescott Joule, William and James Thomson, W. J. M. Rankine — thought in ways that combined the practicality of engineers with the mathematical adroitness of the best natural scientists, and they were better able than their solely Oxford- and Cambridge-trained contemporaries to elucidate the basic laws of heat, and to apply them to the study of cold.

The work of Joule would eventually lead directly to the elimination of caloric and to the "dynamical" theory of heat, which defined heat or cold as a state of motion measurable by its kinetic energy. But that accomplishment did not come easy or quickly. Joule was born on Christmas Eve in 1818, the second son of a wealthy Manchester industrialist. In poor health from age five to twelve, confined to bed, he became a prodigious reader. He emerged from his sickness with a permanent spinal weakness that gave him something of a hunchback and with a personality universally described as unassertive and shy.

Already in this chronicle we have encountered an unusual number of scientists who were sickly in their childhood and who contended throughout life with chronic illnesses. Studying mid-twenti-

eth-century scientists, psychologist Bernice T. Eiduson found a dis-
proportionate number who had been confined to their beds for
large amounts of time by childhood illnesses. During these travails,
they "searched for resources within themselves and became com-
fortable being by themselves"; most turned to reading, and through
reading they developed a bent for intellectual work. Not very good
at sports, unfit by illness to compete in childhood games, they
remained emotionally fragile throughout life, deriving satisfaction
mostly from intense involvement in science.

James Joule and his older brother, Benjamin, who became a noted
musician, were tutored at home and then, under the auspices of the
Manchester Literary and Philosophical Society, by one of England's
most illustrious scientists, John Dalton. A Fellow of the Royal Society, Dalton had produced such milestones as the basic explanation
for atomic weights, and verification that all gases have the same
coefficient of expansion. "It was from his instruction," James Joule
later wrote, "that I first formed a desire to increase my knowledge by
original research." The teenaged Joule shot off a pistol, to measure
the recoil, and burned away his eyebrows; he gave himself and his
friends shocks from electric kites and Leiden jars; he sent electricity
through an old cart-horse and also through a servant girl, recording
her observations on the effects until she passed out.

In 1837 Dalton had a stroke and retired from teaching. Joule was
nineteen and went to work as a manager in the family brewery
between nine and six each day. He spent his other waking hours
performing experiments and writing them up. His first articles were
about electricity; in one, signed only with an initial, he correctly
identified the compound nature of lightning, many years before
photography was able to document it. An 1839 paper was rejected for
publication by the Royal Society, though an abstract was published
in a secondary journal. Many years afterward, Joule told a biographer he had not been surprised by the rejection of his early paper,
because "I could imagine those gentlemen in London sitting round
a table and saying to each other, 'What good can come out of a town

where they dine in the middle of the day?'" A more recent biographer discounts this as Joule's idea of a joke, but it probably contains more than a grain of emotional truth.

Joule continued his experiments, moving from electricity to the study of the heat produced by an electromagnetic engine. He proved that Ohm's law governing the relationship between electric current and resistance held, "whatever the length, thickness, shape or kind of metallic conductor," and that the heating effect of a current is proportional to the resistance of the circuit and to the square of the amount of current. This work did command the attention of the Royal Society, which printed his article about it. But lacking a champion other than Dalton, himself somewhat of an outsider, Joule appears to have been judged as a throwback to the days of amateurs whose contributions only incrementally added to scientific knowledge.

Lyon Playfair, a chemistry professor in Manchester in the 1840s, remembered Joule then as "a man of singular simplicity and earnestness," who once proposed to Playfair that they take a trip together to Niagara Falls, not to enjoy the beauty of the natural wonder "but to ascertain the difference of temperature of the water at the top and bottom of the fall." They would meet over supper to discuss the progress of their research, and Joule also brought Playfair to the brewery, where he had a small laboratory; there, Playfair recalled with perfect hindsight, "I, of course, quickly recognized that my young friend, the brewer, was a great philosopher."

Joule was a philosopher undergoing a change of heart. In 1839 he had confidently predicted that electromagnetic machines would replace steam engines and had tried one out in his brewery expressly for that purpose. In 1841, though, he publicly admitted his inability to perfect an electromagnetic machine that could do better than a Cornish steam engine that raised 1.5 million pounds to a height of 1 foot by the combustion of 1 pound of coal. A third historian of thermodynamics, Crosbie Smith, suggests that Joule's attempt to understand the failure of the electric machine to match the economy of heat engines led him to his most important research, measuring the amount of work obtainable from various amounts of heat.

As frequently happens in science, the scientist's private circumstances changed just when Joule was altering his experimental focus. In 1842 Benjamin Joule married the family governess, disturbing the closeness of the brothers, and in 1843 Joule's father moved the family to a new home and built a laboratory there for his son James. That year, James Joule finished a paper entitled *On the Caloric Effects of Magneto-Electricity, and on the Mechanical Value of Heat,* in which he overturned nearly everything previous researchers thought true about heat. By experiment, he refuted the theory — related to caloric — that what an electric current did was transfer heat from one part of a circuit to another. Joule showed that the current (and only the current) generated the heat, and he argued that heat was not a substance but a *"state of vibration"* that could be "induced by an action of a simply mechanical character." This was actually a revolutionary idea that in another ten years would become the basis for the "dynamical" theory of heat that enabled a proper understanding of cold and how it could be produced.

In 1843 Joule readied this paper for a meeting of the British Association for the Advancement of Science, held in Cork, Ireland, but for various reasons, including his own lack of assertiveness, he was assigned to deliver it to the Chemical section, rather than to the section on Physics. Most listeners ignored it.

This third rejection of a demonstration of a basic principle of thermodynamics — and the articulation of a reasonable new theory of heat — was even less appropriate than the receptions given to Carnot and Mayer, since this seminal paper was written by a man whose work had been previously published by the Royal Society, was composed in proper scientific language that used good mathematics and cited relevant prior research, and was presented to a high-level audience ostensibly capable of appreciating its importance.

Disappointed, Joule kept on, taking heart from the fact that his earlier work on electricity had been confirmed by French and German scientists. He shortly developed an obsession: designing a beautifully structured series of experiments to prove once and for all that heat and mechanical work were equivalent and that the relationship between them was fixed and measurable.

If heat and work were equivalent, Joule reasoned, that would be true even if the heat was generated by something not usually thought of as a heat source. Using a perforated cylinder, he pushed water through its pin-sized holes and found that this slightly raised the temperature of the water; computing the foot-pounds involved in pushing the water through, he came up with a figure that was close to those he had deduced in other experiments.

Since heat was also generated by compressing a gas, Joule tried to measure the work associated with that, placing a copper vessel inside a calorimeter (a device used to measure quantities of heat) filled with water, introducing compressed gas into the vessel, and measuring the resulting rise in the water temperature. The results further validated his earlier experiments.

As had Boyle two centuries earlier, Joule realized there would be objections to his results from people who still believed in an outmoded theory, so he carefully designed further experiments to withstand their objections. Among the calorist critics Joule hoped to refute was Clapeyron. Joule was incensed by Clapeyron's contention, derived from Carnot, that in the transfer of heat from furnace to boiler in a steam engine, large amounts of heat were lost. "Believing that the power to destroy belongs to the Creator alone," Joule wrote, "I entirely coincide with . . . the opinion that any theory which, when carried out, demands the annihilation of force, is necessarily erroneous." This was a religious objection to the idea — actually, the same objection that had troubled Carnot. But Joule knew that his experiments, even if they did not prove or disprove that "force" could be annihilated, did prove some things of equal importance: that heat and work were measurable, and that when air expanded without doing work, no heat was lost or transferred. The implication of the last result was clear: caloric had nothing to do with heat.

Fifty years later, the consummate experimenter James Dewar would deem Joule's success at proving his contentions about heat and cold through experiment "simply astonishing," given that such nimble minds as von Rumford's and Davy's had worked in the same area but had entirely missed the proper understanding. Joule's article put science on the verge of casting away the mechanistic worldview that depended on vague substances such as caloric and of substituting a more mature concept in which heat was the product of molecular motion. Shortly, such understandings would lead to the concept of energy, the transformations of which would explain the production of heat and cold, and also (though tangential to this story) the character and propagation of light, sound, electricity, and magnetic fields.

In 1844, though, when a distinguished older scientist sent Joule's article to the Royal Society for consideration by the *Philosophical Transactions,* it was turned down. At that moment in time, it appeared that the scientific establishments of the three leading experimentalist countries — France, Germany, and Great Britain — would continue to refuse to acknowledge the new understandings of heat, and the lack of heat known as cold, and thereby serve to keep the country of the cold unknown and unexplored.

Through Heat to Cold

THE MAN WHO WOULD DO THE MOST to complete the conquest of cold by reaching an understanding of heat, William Thomson, was twenty-one years old in 1845, the author of a half-dozen interesting papers on physics, the possessor of a recent advanced degree in that field, and was working in Paris at the laboratory of one of France's most respected scientists, Victor Regnault. Thomson's brilliance, already widely acknowledged by his peers, was manifest in everything he did and touched: sitting in a Paris café, he wrote to his friend Gabriel Stokes that he noted in "a cup of thick chocolate au lait" an instance of elasticity in an incompressible liquid when he stirred the liquid with a spoon, then removed the spoon, producing "oscillations" and "eddies." He saw connections everywhere. Examining rival theories of electricity, Thomson tried to synthesize them, and in the attempt he managed to disprove the notion of electricity as a weightless, "imponderable" fluid. But he continued to believe that heat depended on that other imponderable fluid, caloric, as did Regnault, who had devoted his life to the measurement of heat. However, while Regnault would never get beyond mere measurements and his attachments to old concepts such as caloric, the young Thomson would soon be able to leave both behind.

His break with the past started in Paris at the moment when colleagues at Regnault's laboratory introduced him to Clapeyron's exegesis of Carnot. This single article began to deeply affect Thomson's thinking about heat and cold, so much so that he roamed the bookstores of the city seeking a copy of Carnot's *Réflexions*, but could not find a copy to buy, nor even one to study in a library.

The second son of a professor of engineering at the University of Glasgow, William had matriculated there at the age of ten, partly to accompany his older brother, James — the boys' mother had died when William was six. The brothers' minds and interests were complementary, the future engineer James often playing the role of what a later colleague dubbed "the philosopher, who plagued his pragmatical brother." Both held firm beliefs in God, and in God as the first cause of the physical universe, though James was more doctrinaire in religious adherence.

William Thomson was a person so bursting with ideas that he seldom restrained himself enough to allow a partner in a conversation to finish a thought before voicing another of his own. At Glasgow, he studied the great French mathematicians so intently that at age seventeen he produced a paper, *On Fourier's Expansions of Functions in Trigonometrical Series.* As though to prove that was no flash in the pan, when he transferred to Cambridge the next year he wrote *On the Uniform Motion of Heat . . . and Its Connection with the Mathematical Theory of Electricity,* drawing mathematical parallels between the operations of heat and electricity to help explain how heat reaches equilibrium; the certitude and depth of understanding of the eighteen-year-old were breathtaking. By 1845, when he graduated from Cambridge in the position of second "wrangler" — the first prize having gone to a man who paid more attention to his studies while Thomson busied himself writing two additional, highly praised papers — William was already viewed as belonging in the top rank of scientific intellects, and only by age a junior member of the scientific establishment.

It was with his bent toward synthesis and his growing fascination with Carnot in mind that Thomson, along with Stokes, attended a session of the British Association for the Advancement of Science

(BAAS) in June 1847 and heard something that affected him — and the history of work on the cold — even more than his introduction to Carnot: the clarion call of James Joule.

The "British Ass" had begun in 1830 in response to a scathing attack on the scientific establishment that charged that British science was inadequate, poorly taught in universities, and ignored by the government. Seventeen years after its birth, the BAAS had become the most important annual colloquium of serious scientists. They swelled the lists of presenters in attendance at the June 1847 session to the point that the chairman of the Mathematics and Physics section, on the grounds that it was the end of a long day, refused to permit Joule to make a full presentation. However, in the few minutes allotted, Joule exhibited the paddle wheel he employed to measure how fluid friction generated heat.

The genius of science often expresses itself when someone recognizes as important a fact or an idea that others pass over as inconsequential. Such a fact was Joule's experimentally proved data that a paddle wheel moving through water produces heat. This fact was startling — first, because it demonstrated that heat was generated by a method no one had ever thought about in that connection, and second, because it undermined the caloric explanation of heat. Joule believed that his work on the paddle wheel would have sunk into obscurity had not a man in the back of the hall stood up and asked penetrating questions. Thomson remembered the occasion differently; he had wanted to rise with an objection — that Joule was contradicting Carnot — but realized, as he later wrote, that "Joule had a great truth and a great discovery, and a most important measurement to bring forward." So, according to Thomson, he and Stokes waited until the meeting ended to approach Joule and start a discussion.

"Joule is I am sure wrong in many of his ideas, but he seems to have discovered some facts of extreme importance, as for instance that heat is developed by the friction of fluids in motion," William wrote to brother James, enclosing two papers the Manchester brewer had given him, which he predicted would "astonish" James. His brother, who had along with William become a professor at Glasgow, agreed with that assessment, finding a key inconsistency in one of Joule's conclusions but praising Joule's experimentally derived data, and predicting that Joule's ideas would "unsettle" anyone who had become convinced by Carnot as interpreted by Clapeyron.

William Thomson was completing work on a new thermometric scale inspired by Carnot, whose original book he had still not managed to locate. The Réaumur, Fahrenheit, and centigrade scales were merely "arbitrary series of numbered points of reference," Thomson argued, because each degree did not represent precisely the same amount of work. He constructed an "absolute" scale using Carnot's notion that a given amount of heat passing between two temperatures can produce only a particular amount of work. On Thomson's scale each one-degree increment represented an amount of work equal to that of every other one-degree increment. It had no fixed references, such as the boiling point of water, and the only zero was absolute zero. In the previous hundred years a dozen different numbers had been suggested as the value of absolute zero, and these guesses varied by as much as 1,000 degrees Fahrenheit. Thomson was happy to have available the figure for absolute zero determined by his French mentor, Victor Regnault, who — with his usual thoroughness — had averaged the calculations done by four methods to come up with the figure of $-272.75°C$.* Years later, after Thomson was awarded the title of Lord Kelvin, his absolute scale would become known as the Kelvin scale, which would be used in the many concerted attempts to understand the region of utmost cold.

Today Kelvin is generally known for that scale, in the short form of science history — but it was actually a lateral matter emerging from his work on the scale that more markedly advanced scientific understanding of heat and especially of cold: his interaction with

* The figure for absolute zero was eventually refined to $-273.15°C$.

James Joule. In a footnote to the absolute-scale article, Thomson paid homage to Joule's "very remarkable discoveries," but he also contended that Joule had not proved the interconvertibility of heat and work. He ignored entirely Joule's idea that caloric might not exist. Professional courtesy would have insisted that Thomson send this paper to Joule before presenting it, so Joule could have a chance to rebut the remarks about his work, but Thomson did not send it. There is something very sad about his callous treatment of Joule in this initial period; Stokes corresponded with Joule easily, as with a comrade in arms, but Thomson seems to have been unwilling to permit a brewer from Manchester, who had not trained at Cambridge or Oxford, to have thoughts on his own level, even though he could not help being troubled by Joule's ideas. When Joule read Thomson's article in a journal, he fired off a private letter to Thomson that praised him but would not let him off the hook. Joule reasserted that he had proved the interconvertibility of heat and work, as opposed to Carnot's supposition that heat could perhaps be entirely annihilated.

Agonized, Thomson responded with a nineteen-page missive, confessing his inability to answer Joule's objections, especially Joule's objection to Carnot, but saying he also believed that Carnot was correct and hoped to reconcile the two at some point in time.

He tried hard to do so — after a friend finally gave him a copy of the *Réflexions* — in an 1849 article that was a total analysis of Carnot. In it, Thomson restated a Carnot proposition, writing that "nothing can be lost in the operations of nature — no energy can be destroyed." But then Thomson went on to ask a key, Joule-inspired question: When (as Carnot contended) "thermal agency" was spent in conducting heat from a hot to a cold body, "what becomes of the mechanical effect that it might [otherwise] produce?" Thomson had as yet no answer for that question.

William's brother James tried to resolve for him what still seemed like a contradiction between Carnot and Joule. Long ago Robert Boyle had proved that water expands with "terrific" force when it becomes ice; if Carnot was correct, James argued, then that physical change from water to ice — from hotter to colder — ought to produce some useful work. But since the temperature of both water at the freezing point and ice at the melting point was the same, 0°C, there seemed to be *no* work done during the change. James showed that when pressure lowered the melting point of ice, making the conversion from water to ice into an action accompanied by a fall in temperature, work was indeed done. William embraced this result as a vindication of Carnot.

Even with James's help, William Thomson in 1849 could not reconcile Carnot and Joule, and that may have stalled his train of thought about heat for a while. His research was also affected by the sudden death of his father — which, he wrote to a friend, was "of a single event, the greatest grief of my life," so dreadful that he feared it would break up the family, which it shortly did. Of course, even when William put aside heat for two years, he still managed to write a major paper on magnetism.

During those two years, Rudolf Clausius of Germany found the way to reconcile Carnot and Joule, thus opening the door for a great deal of further research into heat and cold. In 1850 Clausius was twenty-eight, and his doctorate was newly minted. During his youth in the Prussian part of Poland he had studied music, coming late to mathematics and physics — and perhaps this was the essential difference between his approach and that of Thomson: Clausius's vision was not obscured by the same Idols of the Theatre and of the Tribe that had held Thomson in thrall. Clausius had not been overly steeped in the long traditions of French mathematics and caloric theory, and so he could reach conclusions that Thomson could not yet accept. Also, Clausius drew on relatively recent writing about heat: the opus of Mayer, by this time so addled by rejection and mental illness that he had stopped writing papers; that of another medical doctor, Hermann von Helmholtz; that of Joule, whose work Helmholtz had relied upon but disparaged; and the work of Regnault, Clapeyron, and Thomson himself. The only thing

Clausius had not read was Carnot's book, since he could not find a copy either.

In Clausius's view, in an engine that produced work, two things happened simultaneously: some of the heat was converted (Joule's notion),* while another portion of the heat was simply transferred from the hotter body to the colder one (Carnot's notion). In other words, the ideas of Joule and Carnot were not mutually exclusive; their theories represented two events occurring at the same time, so there was no real contradiction between them. Clausius then deduced two laws of thermodynamics: the first, that the total amount of energy is always conserved, even when heat seems to disappear (because it is simply being converted to other forms of energy); and the second, that the general tendency in nature is for heat to always flow from hot to cold, not the reverse. Clausius proved the second law by showing that the opposite of it was false. He argued that if heat was able to flow from cold to hot "without any expenditure of force or any other change," then one could devise a perpetual-motion machine that would continually pour energy back and forth from hotter to colder to hotter bodies — and therefore, since everyone knew that perpetual motion was impossible, the notion of irreversible flow had to be true.

In his paper, Clausius cited the primacy of Mayer and Helmholtz, to the detriment of Joule, in an ongoing skirmish — it did not rise to the level of a battle — more based on national pride than on primacy of ideas. German and British scientific establishments had long been rivals, with most of the laurels for discoveries going to the British; but with German science coming into its maturity, Clausius wanted to refer to German antecedents. It was also payback, as Joule and Thomson appeared to have ignored Mayer and in general to have slighted German work. Actually, Joule had first been alerted only in 1848 to the possible importance of Mayer's early papers, and since then he had struggled to personally translate them into Eng-

---

* Also made forcefully by Helmholtz in 1847.

lish. Thomson, too, had never read Mayer, but he would shortly go out of his way to cultivate the friendship of Helmholtz. Also, as historians of science confirm, the interconvertibility of heat and work, and the conservation of energy, were better and earlier stated by Joule than by Helmholtz or Mayer.

When Thomson in February 1851 came to his own formulation of the laws of thermodynamics, he claimed to have evolved them mostly by himself, not having learned about Clausius's paper until after completing his own first draft. He seems also to have avoided accepting the insights of the Scottish engineer and physicist W. J. M. Rankine, who had reached much the same conclusions in a paper that also had been published in the interim.

It might seem to the layperson that the first law of thermodynamics, that energy is conserved in the universe, is a pretty straightforward concept and one rather easy to comprehend. But Thomson's path to this and to the second law was tortuous. His seminal article that states the laws made it clear that Thomson's work derived from that of Joule and from the observations of Regnault, and also owed a great deal to his changed understanding of God's relationship to the natural world. Splitting theological hairs, Thomson wrote, in his version of the first law, that "mechanical effect" could be

> *lost to man* irrecoverably though *not lost in the material world,* [because although] no destruction of energy can take place in the material world without an act of power possessed only by the supreme ruler, yet transformations take place which remove irrecoverably from the control of man sources of power which, if the opportunity of turning them to his own account had been made use of, might have been rendered available.

Thomson also made his own, more precise version of the Carnot-based proposition for the second law: "It is impossible, by means of inanimate material agency, to derive mechanical effect from any portion of matter by cooling it below the temperature of the coldest of the surrounding objects." He implied that these laws of thermodynamics had been foreshadowed by the 102nd Psalm's prophecy: that the heavens and earth "all of them shall wax old like a garment" but that God would "endure."

Like Carnot when confronted with the implications of his work, Thomson was shortly thereafter compelled to another, more radical conclusion, one with cosmological implications. From what he called *A Universal Tendency in Nature to the Dissipation of Mechanical Energy*, he drew a consequence that he had never wanted to note but to which science and logic had brought him: that such dissipation meant the sun was "not inexhaustible." This in turn meant that "within a finite period of time past, the earth must have been, and within a finite period of time to come the earth must again be, unfit for the habitation of man as at present constituted," because the earth would be too cold to sustain life. Though Thomson could not bring himself to say so explicitly, his conclusion showed to others that the Bible's timetable for the creation of the earth and the heavens was not factually accurate. And when his conclusion based on the laws of thermodynamics was considered in conjunction with the evidence verifying Charles Darwin's contemporaneous theory of evolution, they together cast serious doubt on the existence of God as defined by the Bible.

Having finally accepted Joule's contentions about the conservation of energy, Thomson then yielded to Joule's entreaties to become a friend and began with him a long series of joint experiments and publications that firmly established the dynamical theory of heat, and that also made possible the next generation's explorations into the nether regions of temperature.

The theory, when fully expressed by Thomson, based on Joule's earlier work, unified the phenomena of heat, electricity, magnetism, and light. All these, it contended, were different forms of energy that were convertible into one another, and the relationship between forms of energy could be expressed by such numerical constants as Joule's mechanical equivalent of heat.

Since Joule had proved the existence of one constant, there must be others, and Thomson joined with Joule to establish them. Thomson suggested the specific experiments that Joule designed and conducted, discussing most of the details with Thomson in advance of the trials through exchanges of letters and the occasional visit by Thomson to Manchester. Delighted at this collaboration, Joule continually deferred to and attempted to accommodate Thomson, except where experimental data would not permit him to agree. Simultaneously with the joint series, the two colleagues continued their own studies, which were published under each individual's name but which were influenced by both men. Two of the experiments directly concerned the generation of cold and would provide the key to later mastery of cold.

In an early letter that Thomson went back and reread, Joule had brought to Thomson's attention the 1834 work of Peltier on thermoelectrics, in which heat or cold could be produced by electricity. Peltier had shown that heat could be either liberated or absorbed when an electric current flowed across two conductors made of different materials. Neither the shape nor the size of the conductors seemed to have anything to do with this "Peltier effect." Joule suspected that thermoelectric effects had to do with the interconvertibility of thermal energy and electrical energy. With a mind far nimbler than Peltier's, and using Joule's hunch, Thomson quickly showed additional reversible thermal effects occurring in thermoelectric circuits, and that the magnitude and direction of what came to be called "Thomson effects" depended on the composition and temperature of the conductor. If there was a difference of 1 degree Kelvin between the temperature at one end of the conductor and the other, when the current moved along in the same direction as the temperature gradient, from hotter to colder, it could produce heat, but when the current coursed *against* the temperature gradient, it could absorb heat, thereby producing cold. Previously considered no more than a curiosity, cold produced by electricity (using Thomson's findings) would by century's end lead to the construction of effective thermoelectric generators and refrigerators.

Joule and Thomson collaborated directly on a second method of producing cold, related to a phenomenon noted by many scientists, that some cooling resulted when gases were released from pressure. Joule had verified this in his two-vessel experiments. Thomson proposed making important alterations to the apparatus to measure this gas-expansion cooling. Joule eagerly agreed.*

At Thomson's suggestion, Joule replaced the copper vessels with long coils of metal piping, and he considerably narrowed the connector between the two coils, to try to prove that the cooling effect of the pressurized air as it expanded into the lower-pressure arena would be offset somewhat by the heating effect produced at the nozzle due to friction. The first experiments on this were inconclusive, so Joule tinkered further with the apparatus. A letter to Thomson written during this period displays Joule's confabulation of personal and scientific enterprises in the friendship:

> The expected stranger arrived safely into the world yesterday morning. It is a little girl, very healthy and strong . . . [and] if, as I hope, you will make it convenient to be at the christening and stand godfather, we might at the same time endeavour to settle the question of heat and cold from air rushing through an orifice. Using a plug of guttapercha with a small hole I find the air to be cooled from 63 to 61½ when rushing at a pressure of 4 atmospheres.

---

* Their initial idea was to test a hypothesis put forth by Mayer, with which both Joule and Thomson disagreed. By this time, Mayer had gained some respect as a theorist, and, after having spent the previous five years doing nothing more mentally strenuous than cultivating his garden in Heilbronn, he had recovered his sanity; he also had won a new champion in Great Britain, John Tyndall of the Royal Institute, who began to tout the insights of Mayer and to belittle the contributions of Joule. Tyndall disliked Joule for having caught and exposed an important error in one of Tyndall's papers, which embarrassed the flamboyant Tyndall. Eventually, this contretemps would provoke a battle over who had been the first to announce the conservation of energy, a battle in which Thomson came galloping to the defense of Joule, to the latter's great satisfaction.

Thomson was busy — not only with his own work but also with pursuing a woman who would later become his wife — and did not attend the christening, or for some months offer any cogent work suggestions to Joule, though he did write a bit disparagingly to brother James of Joule's request that he stand godfather to the child. A more sympathetic friend of Joule's attended the ceremony as Thomson's surrogate.

Joule's restless mind further refined the apparatus, in the direction of smaller and smaller flow passages, until he was satisfied with a nozzle that fit the definition of a "porous plug." That did the trick. Testing many gases, Joule and Thomson found that air, carbon dioxide, oxygen, and nitrogen became colder during the expansion but that hydrogen became hotter. Experimenting further and graphing the results, they discovered a range of "maximum inversion temperatures." At these temperatures, precooled gases expanded with the greatest additional cooling effect. What shortly became known as "Joule-Thomson expansion" — the expansion of pressurized, precooled gases through a porous plug into a lower-pressure vessel, producing a significant decrease in temperature — became the basis for many subsequent efforts in refrigeration, even those used today. We will encounter it as a key concept in the last stages of the drive toward absolute zero.

In the early 1860s, Thomson and Joule's new elucidation of what could produce cold came to influence the theory of heat. Rudolf Clausius found in Thomson's paper on thermoelectricity an important clue relating to an attribute of matter that deeply intrigued him. For years Clausius had been wondering what could explain or measure the apparently universal tendency toward dissipation. Thomson demonstrated in his article on thermoelectricity that materials possessed some internal energy and postulated that it was somehow used for molecular cohesion. Clausius had touched on a similar concept in his 1850 paper, but it was only after Thomson had produced a sort of experimental verification of internal energy that

Clausius pounced on the idea as though it were the Rosetta Stone that could explain what had previously been right before his eyes but had been incomprehensible.

There were not two types of transformation of energy, Clausius wrote in 1865, there were three. Along with mechanical energy being transformed into heat, and heat being transferred from a hotter body to a colder body, there was the transformation that took place when the constituent molecules of a material were rearranged. From this notion, and from the accepted fact that the change from a solid to a liquid, and from a liquid to a gas, involved work or heat, he derived the concept of *disgregation*, the degree of dispersion of the molecules of a body. The disgregation of a solid was low, that of a liquid higher, and that of a gas higher still.

Clausius argued that when a gas was expanded but no work was performed, a transformational change in its energy condition still took place — an increase in its disgregation. To explain this further, Clausius introduced the term *entropy*, a measure of the unavailable energy in a closed system, or a measure of the bias in nature toward dissipation. The greater the disgregation, the greater the entropy.

Building on the work of dozens of investigators over forty years, Clausius finally concluded that the "fundamental laws of the universe which correspond to the two fundamental theorems of the mechanical theory of heat" were "1) The energy of the universe is constant; 2) The entropy of the universe tends to a maximum."

This ultimate, concise, eloquent expression of the forms of energy eviscerated what historian of thermodynamics Donald Cardwell called the "balanced, symmetrical, self-perpetuating universe" of Boyle and Newton, substituting a glimpse of something wholly modern, stripped of theological benevolence, and thoroughly disquieting: "a universe tending inexorably to doom, to the atrophy of a 'heat death,' in which no energy at all will be available although none will have been destroyed; and the complementary condition is that the entropy of the universe will be at its maximum."

In other words, everything will come to rest, presumably at the temperature known as absolute zero. Within a half century of Clausius's pronouncement, the concept of entropy would provide the key to his intellectual heir, Walther Nernst, for refining the understanding of entropy in a way that would allow twentieth-century experimenters to reach to within a tantalizing two-billionths of a degree of absolute zero.

Of Explosions and
Mysterious Mists

I N 1865, THE YEAR OF Rudolf Clausius's seminal paper, his former student Carl Linde began working for a locomotive manufacturer and at the same time helped found the Munich Polytechnische Schule, the first of its kind in Bavaria. Linde later recalled that he had been expected to follow in his father's footsteps and become a minister, but an early infatuation led him to study the power of machines rather than the power of God. Learning physics from Clausius before becoming an engineer, Linde never lost his respect for theory. In 1870 a contest sponsored by the Mineral Oil Society of Halle caught his eye: the challenge was to design a system to maintain 25 tons of paraffin for as long as a year at a temperature of $-5°C$, achieved through artificial means.

Linde addressed the problem as a student of Clausius. He read all he could about the several extant refrigeration systems, including the Carré absorption machinery, which then dominated the field, having found success in the United States as well as in France. Then he subjected the systems to thermodynamic analysis. The one designed by the Geneva-based chemist Raoul-Pierre Pictet was the most efficient, a vapor-compression system that used sulfur dioxide as the cooling medium; it functioned at much lower pressures than its competitors, but the sulfur dioxide sometimes made contact with water and became the very corrosive sulfuric acid, which ate away the metal of the machinery. Linde found that the other systems were based on principles that did not take advantage of what thermodynamics taught about the conservation of energy. So he designed a thermodynamically sound system of his own, without sulfur dioxide, along the lines of a Carnot cycle achieved through vapor compression.

Mechanical Effects of Extracting Heat at Low Temperatures, his article detailing all this, appeared in a new and relatively obscure Bavarian trade journal, where it was noted by the director of the largest Austrian brewery company, who commissioned Linde to design a refrigeration system for a new brewery. Linde-designed refrigerators were so much better than the Carré- and Pictet-designed machines that within a few years his units had replaced the older ones, first in breweries and then in other industrial processes that required cooling, until there were more than a thousand Linde machines at work in factories all over Europe.

Artificially produced refrigeration has been the least noted of the three technological breakthroughs of great significance to the growth of cities that came to the fore between the 1860s and the 1880s. More emphasis has been given to the role played by the elevator and by the varying means of communication, first the telegraph and later the telephone. The elevator permitted buildings to be erected higher than the half-dozen stories a worker or resident could comfortably climb; telegraphs and telephones enabled companies to locate managerial and sales headquarters at a distance from the ultimate consumers of goods and services. Refrigeration had equal impact, allowing the establishment of larger populations farther than ever from the sources of their food supplies. These innovations helped consolidate the results of the Industrial Revolution, and after their introduction, the populations of major cities doubled each quarter century, first in the United States — where the technologies took hold earlier than they did in older countries — and then elsewhere in the world.

A spate of fantastic literature also began to appear at this time; in books such as Jules Verne's *Paris in the Twentieth Century,* set in 1960, indoor climate control was mentioned, though its wonders were not fully explored. From the mid-nineteenth century on, most visions of technologically rich futures included predictions of control over indoor and sometimes outdoor temperature.

In addition to flocking to cities for jobs, Americans also became urbanites in the latter part of the nineteenth century because there seemed to be fewer hospitable open spaces into which an exploding population could expand. Large areas of the United States were too hot during many months of the year to sustain colonies of human beings; these included the Southwest and parts of the Southeast, with their tropical and semitropical climates, deserts and swamps. Looked at in retrospect, the principal limitation on people settling in those areas was the lack of air conditioning and home refrigeration.

In the second half of the nineteenth century, the use of cold in the home became an index of civilization. In New York, 45 percent of the population kept provisions in natural-ice home refrigerators. It was said in this period that if all the natural-ice storage facilities along the Hudson River in New York State were grouped together, they would account for 7 miles of its length. Consumption of ice in New York rose steadily from the 100,000-tons-per-year level of 1860 toward a million tons annually in 1880. But while the per capita use of ice in large American cities climbed to two-thirds of a ton annually, in smaller cities it remained lower, a quarter of a ton per person per year.

When New York apple growers felt competitively squeezed by western growers who shipped their products in by refrigerated railroad car, they hired experts to improve the quality of their own apples. A specialist was hired to help prevent blue mold, a disease affecting oranges, so that California's oranges would be more appealing to New York consumers than oranges from Central and South America. Believing there were not enough good clams to eat on the West Coast, the city fathers of San Francisco ordered a refrigerator carload of eastern bivalves to plant in San Francisco Bay, founding a new industry there. Commenting in 1869 on the first refrigerated railroad-car shipment of strawberries from Chicago to New York, *Scientific American* predicted, "We shall expect to see grapes raised in California and brought over the Pacific Railroad for sale in New York this season."

The desire for refrigeration continued to grow, almost exponentially, but the perils associated with using sulfuric acid, ammonia, ether, and other chemicals in vapor compression and absorption systems remained a constraint on greater use of artificial ice, as did the high costs of manufacturing ice compared with the low costs of what had become a superbly efficient natural-ice industry. Artificial refrigeration finally began to surpass natural-ice refrigeration in the American West and Midwest in the mid-1870s. In the space of a few years, as a result of the introduction of refrigeration, hog production grew 86 percent, and the annual export of American beef (in ice-refrigerated ships) to the British Isles rose from 109,500 pounds to 72 million pounds. Simultaneously, the number of refrigerated railroad cars in the United States skyrocketed from a few thousand to more than 120,000.

Growth of the American railroads and of refrigeration went hand in hand; moreover, the ability conveyed by refrigeration to store food and to transport slaughtered meat in a relatively fresh state led to huge, socially significant increases in the food supply, and to changes in the American social and geographical landscape. "Slaughter of livestock for sale as fresh meat had remained essentially a local industry until a practical refrigerator car was invented," Oscar Anderson's study of the spread of refrigeration in the United States reported. And because refrigeration permitted processing to go on year-round, hog farmers no longer had to sell hogs only at the end of the summer, the traditional moment for sale — and the moment when the market was glutted with harvest-fattened hogs — but could sell them whenever they reached their best weight.

In Great Britain, the Bell family of Glasgow, who wanted to replace the natural-ice storage rooms on trans-Atlantic ships with artificially refrigerated rooms that could make their own ice, sought advice from another Glaswegian, Lord Kelvin, who assisted the engineer J. Coleman in designing what became the Bell-Coleman compressed-air machine, which the Bells used to aid in the transport of meat to the British Isles from as far away as Australia. Because of refrigeration, every region of the world able to produce meat, vegetables, or fruit could now be used as a source for food to sustain people in cities even half a world away. Oranges in winter were no longer a luxury affordable only by kings.

Refrigeration in combination with railroads helped cause the wealth of the United States to begin to flow west, raising the per capita income of workers in the food-packing and transshipment centers of Chicago and Kansas City at the expense of workers in Boston, New York, and Philadelphia. Refrigeration enabled midwestern dairy farmers, whose cost of land was low, to undercut the prices charged for butter and cheese by the dairy farmers of the Northeast. Refrigeration made it possible for St. Louis and Omaha packers to ship dressed beef, mutton, or lamb to market at a lower price per pound than it cost to ship live animals, and when the railroad magnates tried to coerce the packers to pay the same rate for dressed meat as for live animals, the packers built their own refrigerated railcars and forced a compromise.

The enormous jump in demand for meat, accelerated by refrigerated storage and transport, spurred ranchers and the federal government to take over millions of acres in the American West for use in raising cattle. This action brought on the last phase of the centuries-long push by European colonizers to rid America of its native tribes, by forcing to near extinction the buffalo and the Native American tribes whose lives centered on the buffalo. The conventional view of

American history is that it was the "iron horse" that finally killed off the "red man"; but one could with as much justification say that it was the refrigerator.

Cold of the temperature of ice — cold adequate for most tasks of preserving food and medicines, making beer, transporting crops, preventing hospital rooms from overheating — could be produced by ordinary refrigeration. But scientific explorers wanted to journey far beyond the shoreline of the country of the cold into a temperature region dozens of or a hundred degrees below the freezing point of water. This was a region outside the sense experience of warm-blooded human beings, a region so cold that skin and nerves could not even register the intensity of its cold; the only way to measure its grade of cold was through thermometers. To conquer this region scientists would require a more powerful technology. They found it in the liquefaction of gases.

This was a rediscovery, for liquefaction had begun with van Marum and ammonia in 1787, and significant leaps forward had been taken in 1823, when Faraday had liquefied chlorine, and in the early 1830s by Thilorier, who actually went beyond liquefaction to create solid dry ice from carbon dioxide.

In 1838 for an audience at the Royal Institution Faraday demonstrated the remarkably low temperature of $-110°C$, achieved by use of the "Thilorier mixture" of dry ice (carbonic acid), snow, and ether. He might have immediately gone further with liquefaction, using the Thilorier mixture, had he not suffered a mental collapse that friends attributed to the exhaustion of having done enough work in a single year to fill four scientific papers. Modern historians believe Faraday's illness may have been mercury poisoning, a then-unknown malady. Whatever the cause, bad health kept Faraday out of the laboratory until 1845; but as soon as he recovered, the possibilities for achieving lower temperatures by means of the Thilorier mixture induced him to return to liquefaction experiments. So enabled was Faraday by the Thilorier mixture that despite having otherwise primitive equipment — a hand pump to compress the gases, and a laboratory widely regarded as the worst then available to a serious experimenter in London — in a few months in 1845 he liquefied all the known gases, with the exception of six he could not change, and which were dubbed "permanent" gases: oxygen, nitrogen, hydrogen, nitrogen dioxide, methane, and carbon monoxide.

The "permanent" gases were a significant scientific problem worthy of a strong attack by a scientist of Faraday's brilliance, but he seems to have decided in 1845 that he had exhausted the limits of his new tool, and he went no further in liquefaction, instead returning his attention to electricity and magnetism. Less brilliant researchers on the Continent took up the challenge. Recognizing that the weight of seawater produced high pressures, a French physicist named Aimé first compressed nitrogen and oxygen by ordinary means into metal cylinders, then lowered the cylinders into the ocean, to a depth of more than a mile. He recorded pressures of up to 200 atmospheres — about 3,000 pounds per square inch — but no liquefaction of the gases occurred. Johannes Natterer of Vienna, whom one historian calls an "otherwise undistinguished medical man," thought the problem of liquefaction basically simple: if Boyle's law held, all he needed to do was raise the amount of pressure on the gas, which should decrease its volume to the point of liquefaction. So he kept beefing up his apparatus until it was able to exert as much as 50,000 pounds of pressure on nitrogen gas. But even under such pressure, the gas would not liquefy.

Two abler researchers now addressed the problem, each from a different direction. One of the most astute scientists of the time, the Russian Dmitri Ivanovich Mendeleyev, compiler of the periodic table of atomic weights, started from the liquid state, trying to determine precisely at what temperature any liquefied gas could be again induced to become a vapor. That was a logical approach, but it did not prove fruitful. The opposite approach — to determine the conditions required to make a gas become a liquid — was adopted by a Scottish physician living in Belfast, Thomas Andrews.

The eldest son of a Belfast linen merchant, Andrews had a bent for chemistry, and at age seventeen, having exhausted the facilities for chemical study in Glasgow, he searched through several capitals in Europe for a laboratory to work in before making contact in Paris with a young chemist in the process of becoming a distinguished teacher, Jean-Baptiste Dumas. Andrews returned to Ireland in the late 1830s to study medicine and to teach chemistry. He practiced medicine for only a short time, then became absorbed in a series of experiments on the heat produced or consumed during chemical reactions. By 1849 he was the vice president of the new Queen's College in Belfast, its first professor of chemistry, and a Fellow of the Royal Society. He labored for five years on inconclusive experiments to determine the composition and density of ozone gas. But in the early 1860s these led him to his most important work, exploring in a systematic way what no one had adequately charted, the murky region in which gases and liquids transmuted into one another.

Andrews was not a highly innovative thinker — he missed some obvious things about ozone — but he picked the right substance to work with, carbon dioxide, known to change from gas to liquid under only moderate pressure. As he measured the volume of carbon dioxide gas while compressing it by various amounts of pressure and holding it at a constant temperature, he initially found that Boyle's law adequately predicted the lines on his graph. Those lines are called "isotherms" because they describe the relationship of pressure and volume along a line of a single, constant temperature. Until the time of van Marum, all investigators had found smooth isotherm lines. Van Marum had recorded a discontinuity — a change in the character of the line — at the point where pressure converted ammonia gas into a liquid, but he had not followed that experimental red flag by varying the temperature or by creating other isotherms along which to measure. Andrews did both. With better pressure equipment, he pushed carbon dioxide back and forth between gas and liquid, and he recorded precise measurements of its volume and composition at nearly a dozen different temperatures. He discovered that along the isotherms of an entire group of lower temperatures, certain combinations of volume and pressure kept gaseous and liquid carbon dioxide in equilibrium. Exploring further, Andrews found that so long as he kept carbon dioxide gas above its "critical temperature," no matter how much more pressure he applied, the gas would not liquefy.

This was quite a revelation, and it cried out for Andrews to deduce a new law from it. Because while at relatively high temperatures Boyle's law adequately described the inverse-square relationship between volume and pressure, below the critical temperature it did not, and some other law must apply — one Andrews initially shrank from trying to formulate. He was willing to make only the generalization "that there exists for every liquid a temperature at which no amount of pressure is sufficient to retain it in the liquid form." Later, becoming more daring, he articulated an idea that boggled other minds along with his own: gases and liquids were not independent and fixed forms of matter; rather, each substance (like carbon dioxide) existed along a continuum, and under certain pressure and temperature conditions it could become a liquid, a gas, or a solid.

Scientists had been inching up on this realization for the better part of a century, but before Andrews it had not been boldly stated, nor had experimental evidence been produced to support it. Even Andrews did not comprehensively formulate the notion, leaving unanswered many questions about the states of matter. But he did realize the implications of his work for the further exploration of the cold. Andrews predicted, "We may yet live to see, or at least we may feel with some confidence that those who come after us will see, such bodies as hydrogen and oxygen in the liquid, perhaps even in the solid state."

Andrews's continuum idea was greeted with some interest by his fellow British scientists, but with intense excitement by the Dutch physicist Johannes Diderik van der Waals. The son of a carpenter, van der Waals had struggled through his twenties to find a vocation;

he became a grade-school teacher, then a high-school teacher, then a headmaster in The Hague. These positions did not satisfy him, and while he continued his supervisory work, he also studied physics at Leiden in the late 1860s. It was there that he encountered the work of Andrews and of van Marum, and of Boyle before them.

While Andrews had managed to disprove Boyle's law, he had not successfully substituted a new mathematical explanation that took into account Boyle's results, and van Marum's, and his own. Van der Waals determined to craft an explanation that would encompass all their disparate results, an explanation of the interaction of pressure, volume, and temperature that would even settle such lateral matters as the disparity that Joule and Thomson had found between the cooling effect produced by expansion of nitrogen and oxygen and the heating effect of expansion of hydrogen. He did so in his 1873 doctoral thesis, *Over de Continuiteit van den Gasen Vloeistoftoestand* (On the Continuity of the Gaseous and Liquid State), and in a later, follow-up paper.

At the heart of his explanation was the idea of molecular attraction. Van der Waals replaced Boyle's oversimplified equation with a pressure-volume-temperature equation that factored in an additional variable, molecular density. He expressed mathematically the way in which the molecules of a gas were less densely packed than they were in a liquid. The word *gas* originally derived from the Greek word *kaos,* or "disorder," and the notion that the molecules of a gas were less orderly and dense than those of a liquid made intuitive as well as mathematical sense. When graphed, the temperature line described by van der Waals's equation looked very similar to the graphed isotherm line that Andrews had charted, except in the "equilibrium" area where gas and liquid coexisted. In that section, where Andrews's line was straight, van der Waals's line was subtler, curving down, and up, and down again, along the same pressure axis.

Those subtle, reciprocal curves were another way of showing that in the equilibrium area, for any given pressure, a particular temperature could be produced by three different volumes. Van der Waals's work opened up all sorts of possibilities for future research. What would soon prove to be useful was the idea that a gas-liquid mixture contained molecules both of lower density and higher temperature (in the gaseous portion) and of higher density and lower temperature (in the liquid portion). If one could remove from the mixture the higher-temperature, lower-density gas molecules, leaving behind only the lower-temperature, higher-density molecules of the liquid, that liquid would cool off very rapidly — as though by forced evaporation.

An indication of the importance of van der Waals's thesis was that men deeply interested in thermodynamics and the physics of gases, such as James Clerk Maxwell in Great Britain, tried to learn Dutch just to read it.

The stage was now set for a grand assault on the "permanent" gases. Although initially undertaken to accomplish the liquefactions that Faraday had once thought were beyond the capacities of science, this assault had ramifications far beyond that important goal. For it led to a campaign to reach absolute zero, the lowest temperature imaginable, and in doing so revealed new and unexpected aspects of the nature of matter — and, not incidentally, opened up the country of the cold for the extraction of practical applications that would change society in far greater ways than mere refrigeration had already begun to accomplish.

Among the many scientists in Germany, Switzerland, Great Britain, the Netherlands, and France who took up the challenge of liquefying the so-called permanent gases was Louis-Paul Cailletet, a graduate of the school of mines who was employed at a smelting works owned by his father. Occasionally Cailletet would send a communication to the Académie des Sciences, which would sometimes publish a brief item noting what he had done, for example, in tracing the permeability of metals to gases, but would not really comment on it.

Around 1870, Cailletet moved back to his birthplace, Châtillon-sur-Seine, and set up a private laboratory. Châtillon was close to Paris, and the Académie regarded residence in Paris as prerequisite to being elected to membership. Cailletet had been turned down in his first attempts to seek membership. He was in good company. In the fifty years since the Académie had given short shrift to Carnot, it had grown even more encumbered by tradition and drunk with its own scientific importance, which remained considerable. Louis Pasteur spent more than twenty years in unsuccessful attempts to get himself elected to membership before ascending to the pantheon. Joseph Valentin Boussinesq, an expert on the diffusion of heat, had applied unsuccessfully five times between 1868 and 1873. Even political upheaval — the loss of Alsace and Lorraine to Prussia, the end of the empire of Napoleon III, the establishment of the Third Republic, the evisceration of the Paris Commune uprising, the formal adoption of a republican constitution in 1875 — in no way undermined the stature of the Académie; rather, these events emphasized its continuity and importance as a pillar of French life. To have a report of one's work read and discussed in the Académie's august confines was more than ever essential to establishing a scientific reputation and, not incidentally, a basis on which to claim priority for a patent.

In the fall of 1877, there had been no deaths of full members, and so there was no possibility of the forty-five-year-old Cailletet being selected to succeed anyone. The prospect of becoming a corresponding member, a junior position, was held out to him. Determined to obtain it, he would do nothing to jeopardize the delicate process of accumulating enough positive votes in committee to recommend him. This meant not showing off, or scheduling the report of an important discovery at a time when it might be viewed as attempting to influence the committee's decision. There were the requisite rounds of calls to make to senior members, who would, of course, be certain not to give any hints of their approval or disapproval of the applicant. There were also alliances whose aid must be enlisted or their enmity avoided — the group of students of a single influential teacher, the graduates of a school such as the École Polytechnique (whose members made up 25 percent of the academicians), the adherents of a particular theory, or those who disliked researchers who worked on practical problems or for industrial concerns.

As with many researchers, Cailletet had been captivated by the challenge of doing what Faraday had thought impossible and spurred to action by the observations of Andrews and the theory of van der Waals, both published in the early 1870s. It had become clear that the right combination of pressure and lowered temperatures ought to be able to liquefy almost any gas, and the task had devolved to the technological problem of finding or making the right apparatus. Cailletet bought turbines to build up pressure, rigged tubing and glass vessels, and with an assistant set to work in November 1877 on liquefying acetylene. He chose this hydrocarbon, which was not a permanent gas, because its atomic weight predicted that it could be liquefied with 60 atmospheres of pressure — and in recent years it had become possible to generate considerably more pressure than that under controlled laboratory conditions.

In a manner reminiscent of Faraday's work with ammonia, Cailletet's first try resulted in an accident whose products showed him that he had succeeded in his liquefaction attempt. The cooling-to-liquid stage had occurred as the pressure on the compressed gas had been accidentally released. He did another test with a purer batch of the gas and at a critical temperature of 308 K — that is, above the freezing point of water — and with a "critical pressure" of 61.6 atmospheres, he liquefied acetylene. But this gas was not one of the permanent gases, most of which were components of air. To really get the Académie's attention, Cailletet would have to liquefy oxygen.

To do so, he combined recent pressure technology — he needed 300 atmospheres — with refrigerating processes that had been evolving in laboratories since the experiments of William Cullen in 1748. Cailletet surrounded a glass tube holding gaseous oxygen with another that contained evaporated sulfur dioxide, which reduced the temperature of the inner vessel to $-29°C$, all while he was maintaining immense pressure on the apparatus. When he suddenly released the pressure, a mysterious mist and then droplets formed inside the tube containing the oxygen, and slid down its walls, indicating that some of the oxygen had been liquefied. The few drops did not remain long in the liquid state, but the event had occurred. Cailletet guessed that the sudden fall in pressure had pushed the temperature down to about $-200°C$.

This happened on Sunday, December 2, 1877. Cailletet immediately wrote a letter describing his success and his methods to Henri Sainte-Claire Deville, a friend and a full member of the Académie. Then he turned crafty. The Académie's weekly meeting took place at three o'clock on a Monday afternoon; in December, that meant on the 3rd, 10th, 17th, and 24th. Cailletet's potential election as corresponding member was scheduled for December 17. He determined to win election on that date and to announce the liquefaction of oxygen on Christmas Eve. Toward that end, he organized a demonstration of his liquefaction of oxygen at the École Normale on the 16th. His assistant knew assistants in other laboratories, and word spread quickly about the demonstration. It went off as planned and may well have tipped the balance, for on the 17th, Cailletet was elected as a corresponding member of the Académie des Sciences by a vote of 33 to 19. Only after that did he send in his "communication," with the expectation of having it read on the 24th.

Christmas Eve in Paris was always a festive yet solemn time, with an air of anticipation of the next day's celebration. At three in the afternoon, the academicians gathered in the chapel under the gilded dome of the former Collège des Quatre Nations, opposite the Louvre, on the left bank of the Seine. Many wore their distinctive black uniforms with the green embroidery. There was Victor Regnault, who was very ill and expected to die shortly. There were the former students of Jean-Baptiste Dumas: Deville, Charles-Adolphe Wurtz, Auguste Cahors. Each member sat in his designated place, the seats arranged in an ellipsoid form surrounding the central lectern, and each having a view of the area that had once been the altar of a chapel and was now the site of the desks of the officers and permanent secretary. Beyond, on three sides, were rows of benches for the public, with some reserved for journalists. The desk of the permanent secretary was piled high with letters, articles, and books addressed to the Académie by dozens of applicants, many of whom were unknown to the academicians. On a December afternoon, the chamber would have been illuminated by gas lighting, which had replaced candlelight only two years earlier. Heating columns in the four corners of the room supported busts of a few scientists such as Laplace, and on the upper part of the back wall hung portraits of others, including Lavoisier.

The significance of this particular occasion was pointed out by Dumas, the permanent secretary and the principal champion of restoring the reputation of Antoine Lavoisier. After Lavoisier had lost his head to the guillotine in 1794, there had been a concerted effort by some academicians to retroactively disparage his work, an effort that gained currency when the theory of caloric was disproved. Through the efforts of Dumas and others, the reputation of the father of French chemistry had been somewhat restored. Now Dumas found it appropriate to silence any lingering doubts about Lavoisier, and to link in the minds of his audience the report to come with the great traditions of French chemistry, by reading from the work of the master a speculative rumination as to what might occur if the earth were suddenly transported into "the very cold regions" of the solar system; there, Lavoisier fantasized,

> the water of our rivers and oceans would be changed into solid mountains. The air, or at least some of its constituents, would cease to remain an invisible gas and would turn into the liquid state. A transformation of this kind would thus produce new liquids of which we as yet have no idea.

Only after this prologue was Cailletet's "communication" read. As expected, there was considerable excitement. The most important

caveat was voiced by a member who argued that the real breakthrough would occur when one could create a larger quantity of
liquefied oxygen and maintain it as a liquid for more than a second
or two.

Then, just as Cailletet was basking in the glory, the secretary
announced that a telegram had been received two days earlier from
Raoul Pictet, the Swiss-born physicist who was already a pioneer in
commercial refrigeration. The telegram said: "Oxygen liquefied today under 320 atmospheres and 140 degrees of cold by combined use
of sulfurous and carbonic acid."

Cailletet was aghast, and even more so when a Pictet letter —
prepared earlier, anticipating the result — was read that detailed Pictet's quite different approach to the liquefaction of oxygen.

Henri Sainte-Claire Deville came to the rescue. As a former student of Dumas's, he had no difficulty being recognized by the chair.
Deville insisted that Cailletet had priority, revealing that on December 3, when he had received Cailletet's letter, he had immediately
taken it to the permanent secretary, who had signed, dated, and
sealed it. The letter was found and examined, and Cailletet was
awarded the laurel of being the first to liquefy oxygen.

A week later, on New Year's Eve, Cailletet was back at the Monday
afternoon session with the news that while others had been busy
celebrating Christmas, he had returned to his laboratory and liquefied the other major component of air, nitrogen.

In a single week, giant steps had been taken inside a far region of
the cold, and closer than anyone had ever come to absolute zero.

Painting the Map of Frigor

N EWS OF THE DECEMBER 1877 Cailletet and Pictet liquefactions of nitrogen and oxygen excited the scientific world, opening up vast possibilities of research in the extreme regions of the cold. Individuals and teams of scientists began drives toward achieving the absolute zero of temperature as well as investigations into the properties of matter that might be revealed by very low temperatures.

For the next thirty-five years, during the Gilded Age at the end of the nineteenth century, and in the early part of the twentieth, a set of scientific rivals explored the contours and characteristics of what they frequently referred to as "the map of Frigor." They used such terminology because they believed themselves embarked on voyages of exploration, with all of the danger, sense of the unknown, and fervid romanticism summoned up by that metaphor. Geographic explorers were thrusting toward the North and South Poles; this band of physicists and chemists were on a similar adventure, aiming toward their own "cold pole," or "Ultima Thule," the absolute zero of temperature. "The arctic regions in physics incite the experimenter as the extreme north and south incite the discoverer," the director of the Dutch laboratory pursuing ultra-low temperatures, Heike Kamerlingh Onnes, said in a speech. James Dewar of the Royal Institution in London, another leading laboratory in the race toward absolute zero, lamented in a lecture that while the geographic quests had enormous appeal to the popular imagination, the scientific quest for absolute zero had much less, but on the other hand (as the London *Times* paraphrased his main point), "the approach to the zero of temperature will open out fields of investigation where matter and energy can be examined under new conditions. . . . In both cases, success may be said to depend upon equipment, persistency, and the selection of the right road." While there was no inherent reason why the North Pole could not be attained, there were "strong grounds for belief that the zero can never be reached, [and] the nearer we get to it the more important physical problems become." Kamerlingh Onnes also pointed out that because it was so difficult to make progress, every step toward absolute zero would yield extremely important scientific data; the more so, the closer the experimenters came to their goal.

In his maturity, James Dewar claimed that his most formative early experience was the long illness brought on by his falling through the ice of a pond in 1852, when he was ten. During the two years it took for this youngest son of a vintner and innkeeper in Kincardine-on-Forth, Scotland, to recover from the rheumatic fever that followed the fall and his rescue — an illness that crippled his limbs, reducing him to moving about on crutches — the village carpenter taught him how to construct violins, as an exercise to strengthen his fingers and arms. Dewar later referred to his violin-making as the source of his manipulative skills in the laboratory. The connection he did not articulate, but which seems equally true, was that his fall through the ice birthed a fascination with cold that informed and directed his most productive years. Between the moment at the close of December 1877 when Dewar learned about Cailletet's liquefaction of oxygen, and the moment in 1911 when Kamerlingh Onnes discovered superconductivity, Dewar was increasingly consumed with reaching absolute zero and discovering the properties of matter in the ultracold environment, so completely consumed that his research became an obsession.

In 1859, then seventeen, Dewar had matriculated at Edinburgh University, residing there with his elder brother, a medical student; to earn a post as a laboratory assistant, as evidence of his dexterity he displayed one of the fiddles he had made. At Edinburgh he won various high prizes in mathematics and natural philosophy, and during his university years he studied with or assisted in the laboratories of several of the most respected physicists and chemists of the day. He designed a brass-and-wood model of what Friedrich Kekulé, the father of structural chemistry, had postulated as the structure of benzene. Dewar's model showed that the actual benzene ring could be any of six different forms; a leading British chemist sent the model to Kekulé, who asked Dewar to spend a summer with him in Ghent, a signal honor. Appointed in 1875 as Jacksonian Professor at Cambridge, Dewar taught there for the next forty years, but he never fulfilled one of the requirements of that post — to discover a cure for gout — and although he rubbed shoulders there with accomplished physicists and chemists, and collaborated on some fine spectroscopic research with colleague George Downing Liveing, he was not at ease at Cambridge. "The crudity of youth was still upon him," a literary friend, Henry Armstrong, later reminisced, "and the free manners of a Scottish university were not those of conventional Cambridge — his sometimes imprecatory style was not thought quite *comme il faut* by the good. No attempt was made to tame him or provide means for the development of his special gift of manipulative skill."

In addition to Dewar being "a terrible pessimist," Armstrong recalled, he was "not great as a teacher," perhaps because his mind was "too original and impatient" and because "he never suffered fools gladly"; moreover, when Dewar had not meticulously prepared himself, his lectures could be "incoherent." Yet when he did carefully prepare lectures, they were "logical" and "fascinating" — characterizations applied to the March 31, 1876, Friday evening discourse that Dewar delivered at the Royal Institution, the success of which sealed his 1877 appointment there as Fullerian Professor of

Chemistry. With his wife, Helen, the thirty-five-year-old Dewar moved his furnishings into a small apartment at the Royal Institution. The Dewars were childless, as every resident couple had been since the inception of the Royal Institution in 1799. Augmenting the apartment were the facility's well-stocked library, a separate Conversation Room in which tea and coffee were served and where members could peruse journals or newspapers from the Newspaper Room, and other amenities. Dewar went to work in the basement laboratories, which John Tyndall, the current director, had had redone in 1872, demolishing — at great emotional cost to himself, Tyndall let it be known — the place where Faraday had made his many discoveries. This basement "studio," Armstrong later recalled, was much to Dewar's liking, full of heroic scientific ghosts and available to be visited at all hours of the day and night. The only thorn in Dewar's side was Tyndall, who continued as director for a time. Tyndall was an insomniac, Dewar suffered from chronic indigestion; sometimes, late at night, both resident geniuses walked the halls; Dewar could occasionally be heard by the staff, speaking to the ghost of Faraday.

Impressed by the important work done by Davy and Faraday on electricity, magnetism, and several other areas in which chemistry and physics overlapped, Dewar resolved to maintain their level of experiment and insight. When in December 1877 Cailletet showed the way to reopen the series of gas-liquefaction experiments that Faraday had abandoned in 1845, Dewar seems to have decided that the gods of scientific inquiry were indicating the direction his future work must take.

Prior to his installation in the Royal Institution, Dewar had been a thoughtful and deft experimentalist; at Albemarle Street he developed into the greatest showman of science, "a combination of the magician with the actor," Henry Armstrong recalled, with "force and individuality comparable with that displayed by the most individual actor of his time, his friend Sir Henry Irving." An equal influence on the theatrics of Dewar's performance was his stage, the amphitheater at the Royal Institution. Behind the twelve-column, thirteen-window exterior, a grand stairway led to the second floor and the relatively compact and steeply angled open semicircle of an amphitheater, with its dark wooden pews and banquettes seating about 750. Good sightlines from every vantage point and fabulous acoustics won it acclaim as among the world's best theaters for the spoken word. Also adding to Dewar's brilliance was an almost perfect audience, the distinguished, educated, moneyed crowd of the Friday night lectures, an audience that possessed enough elementary understanding of science to appreciate his demonstrations. Many guests would come dressed as for the opera.

In his first lecture as a resident, Dewar demonstrated the production of droplets of liquefied oxygen from a Cailletet machine, along with other fascinating low-temperature tricks. Under his hand, solid dry ice boiled in liquid ether, continuously giving off gas bubbles. Forty years earlier, Faraday had done the same thing in the same venue, using dry ice in ether, the Thilorier mixture. After crediting Faraday with having "a mind full of subtle powers of divination into nature's secrets," Dewar went his predecessor one better, demonstrating that the solid's temperature was so far below ordinary freezing that when he dropped it in water, it did not emerge from the water coated with ice.

As text to accompany such presentations, he scoured the literature of his forebears — contained in the institution's library, which he liked to frequent — to find facts and portents about liquefaction. His most trenchant prediction was a long-forgotten, 1802 gem from John Dalton: "There can scarcely be a doubt entertained respecting the reducibility of all elastic fluids of whatever kinds into liquids; and we ought not to despair of effecting it in low temperatures, and by strong pressure exerted upon the unmixed gases." Dewar promised the Friday Nighters that he would continue the Royal Institution's tradition of work in liquefaction and the exploration of low temperatures.

Dewar seemed to have found in the bag of tricks that low-temperature research made possible, and that were less available in other areas of chemistry and physics, a reason for being that science alone could not provide him. To a far greater degree than many equally competent experimenters, he enjoyed performing and being appreciated by lay audiences. Several hundred years earlier, Robert Boyle had acidly described the difference between his own work and that of men such as Cornelis Drebbel, writing that "mountebanks desire to have their discoveries rather admired than understood, [but] I had much rather deserve the thanks of the ingenious, than enjoy the applause of the ignorant." James Dewar sought appreciation from many audiences, and he perhaps pursued low-temperature research rather than other areas because it could provide him with applause from lay audiences as well as from fellow scientists.

For the next several years after his initial Friday night demonstration of the Cailletet process, however, Dewar made little progress, hampered by the same difficulties Cailletet and Pictet encountered in producing more than a droplet at a time of liquefied oxygen and nitrogen. In the meantime, two men at the Jagiellonian University in Kraków, Syzgmunt Florenty von Wróblewski and Karol Stanislaw Olszewski, seized the lead in the liquefaction race.

The year of greatest importance to Polish youth in the mid-nineteenth century had been 1863. Syzgmunt von Wróblewski was eighteen that year; Karol Olszewski, seventeen. Von Wróblewski was the son of a lawyer from the Lithuanian area, and a student at the university at Kiev, then a city under Russian control. Since 1795 Poland had been partitioned, its area split mainly between the Austrian and Russian monarchies; the cause of a free Poland had been in contention afterward, occasioning such outbursts as the proclamation by Karl Marx and Friedrich Engels in 1849 that the liberation of Poland was the most important task facing the workers' movement in Europe. A January 1863 Russian plan to press-gang Polish students into the tsar's army sparked student revolts throughout the land. Von Wróblewski joined that revolt and for his participation was arrested, convicted, and sentenced to hard labor in Siberia.

Karol Olszewski, born only months after his father died in the previous, unsuccessful revolt against the Russians, was naturally also caught up in the ferment of 1863. But before he could make his way to the frontline and commit the sort of offense for which von Wróblewski was sent to Siberia, the Austrian authorities detained him. They released him, unharmed, only after the revolt had been crushed. In 1866 Olszewski entered the venerable Jagiellonian University at Kraków — alma mater of Copernicus — and pursued his interest in chemistry and tinkering, becoming what a later colleague, Tadeusz Estreicher, described as a man of "great practical sense and ability in the construction of machinery." Held back by a lack of funds with which to complete his education and to buy machines and materials for his studies, Olszewski managed after three years to become the assistant to the chemistry professor. After repairing an old Natterer compression machine, he used it to solidify carbon dioxide; though the accomplishment was neither new nor particularly difficult, it established Olszewski as a good laboratory hand. He went to Heidelberg in 1872 to study chemistry under Robert Bunsen, inventor of the Bunsen burner, and returned, doctorate in hand, to become chemistry professor at Kraków in 1876.

Von Wróblewski's path to the Jagiellonian was more circuitous and loftier. In Siberia, interspersed with his bouts of hard labor, he read widely, especially in physics, and constructed for himself what he believed to be an entirely new theory of electricity. Released in a general amnesty in 1869, he was in very poor general health and on the verge of going blind. After two operations and six months in a dark room at a Berlin hospital, he emerged and began to study physics under Helmholtz, though he had been warned not to read or write lest he lose the remainder of his vision. During part of his recuperation, in the Swiss Alps he met Clausius, who encouraged him to continue his studies and to concentrate on things related to thermodynamics. Working variously at Berlin, Munich, and Heidelberg, von Wróblewski received a doctorate for his studies of electricity and did other research on the ways in which gases were absorbed by various substances. Offered a professorship in Japan in 1878, he turned it down in favor of the opportunity of returning to a Polish venue, even though Kraków remained under Austrian control; in exchange for his willingness to come to the Jagiellonian, he received a fellowship to spend the next few years studying in Paris, with forays to Oxford, Cambridge, and London. He wrote to his sponsors that he learned more in five months in England than he had in his previous five years of studies. In Paris he worked with a Cailletet liquefaction apparatus, and he brought one such machine with him when he took up his duties at Kraków in 1882.

It was a moment when the Jagiellonian University, which had been less active in science during the period of the partition of Poland, was again moving toward the forefront of research in a half-dozen varied scientific fields. Very quickly after von Wróblewski arrived in Kraków, the nearly blind theoretical physicist and the mechanically inclined practical chemist, Olszewski, decided to team up to attack the liquefaction of oxygen. The few drops produced by Cailletet were not enough for anyone to experiment with; what was needed was to make and keep a quantity of liquid oxygen at its boiling point, and that was what the Poles set out to accomplish. Though the two men were the same age, von Wróblewski was a full professor but Olszewski was not, von Wróblewski's presence at the Jagiellonian was more highly valued (and remunerated), and his research was well funded, whereas Olszewski had limped along with equipment a quarter century out of date. Today it is not clear which man came up with the idea for an improved way of producing liquefied oxygen; they both later claimed sole credit.

Regardless of whose idea it was, their innovation owed much to the theoretical explanation of the continuity of the gaseous and liquid states provided by van der Waals. Moreover, the basic principle they used — evaporation — dated back to ancient Egypt: the desert denizens regularly cooled foods at night in the open air by putting them under a pan of water and letting the water evaporate, which lowered the temperature of the pan and the food. Van der

Waals had made clear a principle that underlay all evaporation, that even in a liquid-gas mixture, the molecules in gaseous form are less dense than those in liquid form. This principle was the basis of the intuitive leap made by the Poles: their technique drew off the lighter molecules, lowering the temperature of the remaining ones, and resulting in liquefaction. In March–April 1883, utilizing a combination of the methods of Cailletet and Pictet, the new Cailletet pressure apparatus that von Wróblewski had brought back from Paris, and Olszewski's adept handling of machinery, the pair liquefied air, then carbon monoxide and nitrogen, and, finally, oxygen. On April 9, 1883, they triumphantly reported to the Académie des Sciences that a measurable quantity of slightly bluish-color liquid oxygen was "boiling quietly in a test tube" in their laboratory, at a temperature of −180°C.

The dimensions of cold had just become more frigid: from the −90°C of chlorine and methane, to the −140°C of ethylene, down to the −180°C of oxygen. The 90-degree drop from the liquid chlorine of Faraday to the liquid oxygen of Olszewski and von Wróblewski was the equivalent of having lowered the temperature from that of boiling water to that of water so shivering cold that immersion in it would instantly kill a human being. Previously, liquefied oxygen and nitrogen had existed only as short-lived droplets from a mist; now workable quantities of both elements had been obtained.

The man who had received von Wróblewski's telegram in Paris was one of his teachers, and he wrote back conveying personal congratulations from himself and Jean-Baptiste Dumas, the president of the Académie, and to convey the gossip that the Polish feat had occasioned much chagrin that the liquefaction had not been accomplished in Paris. Another effusive letter came to the Poles from Cailletet; von Wróblewski treasured this letter, he said in his return missive to Cailletet, because "it proves a rare greatness of spirit. You express your delight in something that justifiably ought to be your success."

Thus began an era of intense experimentation in liquefaction and on the properties and uses of liquefied gases. Many experimenters would build on the results and techniques jointly developed by von Wróblewski and Olszewski, but not the pair themselves, at least not as a team. Within months of their joint accomplishment, Olszewski and von Wróblewski quarreled and terminated their collaboration. The cause of their split remains obscure to this day; Olszewski's later colleague Estreicher offered perhaps the most balanced and logical assessment of it: "The chief reason . . . was that each of them possessed a strong personality and differed in temperament, which made relations between them difficult; each of them wanted to work in the same direction, but in a different way, and neither would make concessions and be subordinate to the other." During the next five years, Estreicher observed, both the chemist and the physicist continued to conduct more liquefaction experiments, separately and intensively, "as if each of them wished to surpass the other."

The Poles' achievement took the group of questing scientists beyond a mountain peak to a point on the other side, and there now opened up, below, a vista no one had ever before seen: a great valley full of unrecognizable vegetation, rock formations, rivers, and geysers; a valley that invited their descent and exploration but that also promised harsh travel and continual hazard, for at every step they took the temperature fell and the land became stranger and more unlike the warmer territories they had left behind.

To descend further, all the groups adopted the technique Pictet had pioneered and the Poles had improved: the cascade. It was like a series of waterfalls, one beneath the other, the first one gentle but feeding water faster into the next, which in turn fed it still faster into the third, whence it emerged in a boiling roar. In a liquefaction cascade, the temperature of a gas was first lowered by the removal of lighter molecules, by pressure and by cooling, until the gas became a liquid; then that liquefied gas was used to reduce the temperature of a second gas, liquefying it; afterward, the second liquefied gas was used to liquefy a third. Cascades permitted experimenters a wild ride down the mountain, from the −110°C reached by Faraday with the Thilorier mixture, all the way to −210°C, the lowest point beyond liquid oxygen that had yet been reached. Off ahead of them, the explorers could see the next landmark goal, the "critical temperature" at which they should be able to liquefy hydrogen. From the calculations of van der Waals, it was expected to be about −250°C.

Minus two hundred and fifty degrees centigrade! A destination so full of dread and so difficult to attain that they almost despaired of getting there, although it was only 40 degrees centigrade lower than what could currently be reached. Dewar, Kamerlingh Onnes, and other researchers reminded colleagues and lay audiences in speeches and articles that in this territory below the temperature of liquid oxygen, each drop of 10 degrees centigrade was the equivalent of lowering a temperature in the more normal range of 100 degrees centigrade, and much harder to accomplish. There seemed no other way to get there but by expanding the cascade series.

In January 1884 von Wróblewski reported producing a *liquide dynamique* (constantly changing liquid) of hydrogen by cooling the gas with liquid oxygen, then allowing it to expand rapidly, which dissipated the energy and lowered the temperature; but the product was not a quietly boiling liquid in a test tube. Almost immediately, Olszewski reported the same result from his cascades: colorless drops running down the side of a tube. Later in 1884, Dewar told the readers of the *Philosophical Magazine* that Olszewski's work in progress meant that scientists would not have to wait much longer for "an accurate determination of the critical temperature and pressure of hydrogen." As things turned out, this was the last good thing James Dewar would ever have to say about Karol Olszewski.

In the meantime, another laboratory had entered the race, one under the command of Heike Kamerlingh Onnes at the University of Leiden in the Netherlands. Kamerlingh Onnes took up his duties as professor of physics and as chief of the research laboratory in November 1882, at the relatively young age of twenty-nine, and after beating out another serious contender for the position, Wilhelm Conrad Röntgen, who in 1901 would be awarded the first-ever Nobel Prize in physics, for his work on x-rays. Onnes was chosen in part because he was thoroughly Dutch, while Röntgen, although he had lived in Holland since the age of three and been educated at Dutch schools, had been born in Prussia.

Onnes grew up in a home that he later recalled as studious and isolated. His father was a roofing-tile manufacturer in Groningen, and because his parents felt themselves more refined and interested in culture than were the other burghers, yet not cultured enough to mingle with the university professors in the town, he recalled, they had few friends. "Therefore we remained at home, read much, talked about art, and developed ourselves consciously, so to say." In that home, a "deep inner culture" was combined with good manners and "neat and careful dress"; the Kamerlingh Onnes boys' entire mode of existence was "subservient to *one* central purpose: to become *men*." A younger brother became a well-regarded painter; another brother, a high government official. A French colleague would later recollect that Onnes would frequently stagger him by the "immensity of his erudition," particularly his knowledge of such matters as French literature.

In grade school, under the influence of the director, a professor of chemistry at Leiden, Heike developed an interest in science. At the university at Groningen, fellow students later recalled, Onnes would complete his schoolwork almost before they had begun their own, and he won first prize for a scientific essay comparing methods of obtaining the vapor density of gases. A fellowship took him to study with Bunsen and Gustav Kirchhoff; under their influence, he delved more into physics, becoming fascinated with Jean Foucault's pendulum, which led him to a doctoral thesis titled *New Proofs for the Axial Changes of the Earth*. It took him four more years to complete his studies, an interim he spent as a lecturer and laboratory assistant to one of the leading physicists of the Netherlands.

Onnes was so impressive when he defended his thesis in 1879 that the examiners dispensed with the usual custom of asking the candidate to leave the room while they decided his fate and instead, a senior chemist later recalled, "unanimously and without discussion" awarded him his doctorate. The preamble to his thesis, a quote from Helmholtz, became the touchstone of his life's work: "Only that man can experiment with success who has a wide knowledge of theory . . . and only that man can theorize with success who has a great experience in practical work." Three years later, upon the retirement of an older professor of experimental physics at Leiden, Onnes ascended to that chair, and to the leadership of the university's experimental physics laboratory, the only such lab in the Netherlands. In his inaugural lecture, he expressed the wish that he could inscribe above every portal in his laboratory the motto *Door meten tot weten,* "Through measurement to knowledge." He also announced a program of quantitative research "in establishing the universal laws of nature and increasing our insight into the unity of natural phenomena." This was a direct reference to van der Waals's theory expressing the unity of gaseous, liquid, and solid states, known as the law of corresponding states, which, Onnes later wrote, "had a special charm for me." He set out to prove the theory through "the study of the divergences in substances of simple chemical structure with low-critical temperature." He deemed the theory so important that he later had plaster casts made of three-dimensional graphs of its equations.

During most of the rest of their lives, Onnes and van der Waals would meet monthly for private talks about the progress of the work. According to van der Waals, Onnes was "almost passionately driven to examine the merits of insights acquired on Dutch soil." Onnes echoed this estimate of his motivation, later writing that "the desirability of coming a step nearer to the secrets of absolute zero, and the fascination of the struggle against the unsubmissive [gases] in the country where van Marum first liquefied a gas are too strong, to allow the question to be forced away from one's thoughts."

At the outset of Onnes's operations, "only comparatively small means were at my disposal." The government of the Netherlands granted a modest subsidy to the lab, but Onnes could allocate just a portion of it to low-temperature research. Ethylene, an essential ingredient for lowering temperatures of other gases, was "very expensive" to purchase, and so he had to devote part of his lab space and time simply to making it. His lone assistant, who took care of the machinery, often had to abandon the construction of new equipment to repair older pieces, resulting in "intervals of stagnation which sometimes did much harm." Purchasing a Cailletet apparatus, Onnes replicated Cailletet's experiments, then those of Pictet, then those of the Poles, altering and improving the apparatus as he went along:

> It took much time to free all pieces from smaller or greater leaks and defects, to lay perfectly tight packings, to make suitable conduits, to make cocks which do not get fixed by the cold . . . to devise gauge-tubes showing the level of the condensed gas and filtering-apparatuses for protecting the cocks. Much that [later became] an article of trade was not yet known and had consequently to be made, which was very troublesome. And moreover there had to be acquired practice in all sorts of unusual work.

Three years into his research, with the Cailletet machine still not in perfect working order, and when he was still badly lagging behind Dewar and the Poles, Onnes did something no other competitor in this race would do. He began a monthly journal, *Communications from the Physical Laboratory of the University of Leiden*, issued in English, that was remarkable for its openness, its willingness to admit mistakes, and its sense of immediacy. Reading it, other researchers were instantly able to know all the important details of what Kamerlingh Onnes was doing, so they could readily replicate his experiments; this was in stark contrast to Dewar's articles and public demonstrations, which did not really reveal his methods and almost never reported his failures or what he had learned from them.

In 1885 von Wróblewski brought together the passion of his youth, electricity, and the low-temperature investigations of his maturity. Looking into the conductivity of copper wire, he found "extremely remarkable properties" at low temperatures, and in a paper he drew attention to the steady rise in the wire's ability to conduct electricity as the temperature was lowered by liquid nitrogen and other such fluids. It was an early first indication that in the far regions of the country of the cold, the conditions that characterized life at normal temperatures no longer applied.

In March 1888 von Wróblewski was working late one night in his laboratory, alone, his fragile eyesight strained to the utmost by the effort to design a new apparatus with which to attempt the liquefaction of hydrogen. He knocked over a kerosene lamp. The glass shattered and poured onto him a stream of the flaming liquid; for the next three weeks he lingered in a hospital bed, then succumbed to his burns, dying at the age of forty-three.

The death of von Wróblewski was noted and mourned abroad, Estreicher writes, but had no effect on the work of Olszewski. The chemist had been attempting to perfect his own apparatus. Several times the glass tubing exploded, setting back his progress. In 1889 to 1890, he switched to metal containers. "This apparatus constituted the greatest progress in the field of the liquefaction of gases and was a real sensation in the scientific world of those days," Estreicher insisted, contending that Onnes and Dewar later adopted it — a claim that would be hotly disputed.

Dewar himself had only recently returned to low-temperature research after an explosion in his laboratory in 1886 that nearly killed him and that severely injured his associates. The accident was so bad that after 1890 in Great Britain, to prevent fatal mishaps that might come from mixing gases, valve threads on some combustible-gas cylinders were made right-hand, while those on the cylinders of other gases with which they might interact were made left-hand.

In 1892 Onnes and his associates finally perfected the equipment they had been working on for a decade, and for which they had even gone to the length of borrowing from the navy a pump formerly used to fill torpedoes with compressed air. Only then — years after Dewar and Olszewski were doing it routinely — were the Dutch able to produce liquid oxygen in useful quantities and to maintain it for their experiments. The apparatus promptly broke, spoiling everything. But not for long. Onnes appears to have used this crisis as the basis for successfully arguing with the elders of the university, and with the Dutch government, that he must have adequate funds and assistants to achieve any real progress. A year later, he pronounced himself proud of the new, large-scale liquefaction plant in his laboratory, which, he said, would enable him to begin a program of liquefying all the known gases and reaching down to the neighborhood of $-250°C$.

Approaching Christmas 1887, the managers of the Royal Institution told the aged John Tyndall that Dewar, rather than he, would give that year's annual Christmas lecture to children; Tyndall resigned, and the managers appointed Dewar director of the Royal Institution; it was rumored that Dewar insisted the Tyndalls vacate the director's flat by January 1.

A balding, middle-aged man in trim beard, starched collar, and formal black suit, Dewar did magic tricks with liquid oxygen boiling in a tube for the adult Friday Nighters. He extracted a drop of the liquid oxygen and put it on his arm, supposedly to "show that it was in the spheroidal condition," but really to demonstrate that he was not afraid of the cold tiger. He added alcohol to the liquid in the test tube, and the alcohol instantly froze into a solid within the liquid oxygen. He held a lighted taper over the test tube, and the vapor given off by the liquid ignited the taper into flaring flames. Having startled his audience, he then waxed philosophic, telling the men in evening jackets and the ladies in ball gowns that as science neared the projected temperature of liquefied hydrogen, the world would learn how those temperatures grandly altered many properties of matter. He prophesied that at or below the temperature of liquid hydrogen, "molecular motion would probably cease, and what might be called the death of matter would ensue."

He could make these predictions with some confidence because he had seen indications of astounding transformations from a lengthy and detailed series of experiments he conducted with J. A. Fleming, beginning in the late 1880s and lasting for many years thereafter. Fleming and his assistants made the measurements, while Dewar directed what was to be done and interpreted the results. These experiments accomplished more than any others in delineating many of the contours of the map of Frigor. It was an almost unworldly landscape they painted, one whose features had been etched by the fantastic transformative power of seriously low temperatures.

Scientists had long since established that a bit of chilling made it possible for metals to become better conductors of electricity — that it lowered the metals' electrical resistance — but Fleming was unprepared for what happened when coils of iron were plunged into liquid air: after such a bath, the coils registered just one-tenth of the "resistivity" they had at room temperature. Dewar and Fleming found that as temperatures dropped drastically, the changes in resistivity were not uniform; in the vicinity of $-200°C$, metals that at room temperature were good conductors — silver, zinc, gold — were eclipsed in conductivity by those that in normal conditions were not as good — copper, iron, aluminum. This research moved Dewar to dream that at the absolute zero point, if it could ever be attained, "all pure metals would be perfect conductors of electricity. . . . A current of electricity started in a pure metallic circuit would develop no

heat, and therefore undergo no dissipation."* Fleming went so far as
to propose a new definition for absolute zero: "the temperature at
which perfectly pure metals cease to have any electrical resistance."

As Dewar widened his research, he reached out to other collabo-
rators, including Pierre Curie, with whom he studied the effects of
extreme cold on the emanations of radium and on the gases oc-
cluded by radium.

At extremely low temperatures, thermometers made of mercury
or other liquids were freezing solid. Siemens in Germany had constructed a thermometer based on the curve describing the decline in
resistance in a platinum wire, a curve that could by extrapolation
give readings for the temperatures of other materials as they became
ever colder. Dewar and Fleming began to use such "resistance" thermometers.

The ability to measure low temperatures did not help Dewar and
Fleming make sense of the finding that a bath in liquid air made
dramatic changes in the insulating capacity of substances that were
already good insulators at room temperature. Glass, paraffin, and
natural substances such as gutta-percha (a variety of natural rubber)
and ebonite (an even harder rubber) did not lose insulating capacity but rather became even better insulators after being immersed
in liquid air. The experimenters had no explanation for that. Nor
could they come up with reasons for what happened to magnetization under the influence of liquid air or liquid oxygen. Dewar and
Fleming were intrigued to find that while most magnets gained
strength when subjected to intense cold, some did not; moreover,
when pure iron was immersed in liquid oxygen, it afterward required a much greater magnetic field to magnetize it than was
needed under normal conditions. Even mercury at very low temperatures became a good magnet, whereas at room temperature it

---

* It would develop no heat because nothing would resist its passage — resistance is what
produces heat in electrically conductive wires, in, for example, the coils of a conventional
electric toaster when they are activated. And if there was no heat, no energy would be
dissipated.

exhibited virtually no magnetic pull. To help explain that, the experimenters reminded themselves that in the periodic table, mercury was listed as a metal.

In the grip of Frigor, iron, copper, and zinc exhibited enhanced rigidity and greater strength: a coil that at room temperature could support only a pound or two of weight could support three times as much after immersion in liquid oxygen. When balls of various metals taken from the bath were dropped on an anvil, they bounced higher than they normally did, leading the researchers to conclude that lowered temperature produced greater elasticity in metals, and to guess that this might be traceable to increased molecular density of the supercooled metals.

It was a good guess, and its likelihood was spectacularly bolstered by the work of Dewar and Fleming on chemical affinities at the temperature of liquid oxygen. Generations of science students had been startled and delighted by demonstrations involving the violent reactions of some chemicals when brought into the presence of oxygen at room temperature, watching as these substances instantly formed oxides, a process that generated a great deal of chemical heat, sometimes accompanied by sparks and burnings. But when such usually volatile substances as phosphorus, sodium, and potassium were plunged into liquid oxygen, nothing happened: they failed to react. Chemical affinity, the researchers discovered, was all but abolished in the extreme cold. Not only that, but chemical combinations that at room temperature would always generate electricity failed to do so at the temperature of liquid oxygen.

The most curious and unexpected of the experimental findings had to do with the optical properties of materials. Under the extreme cold, substances such as mercuric oxide, normally bright scarlet, faded to light orange, while white-colored substances intensified their whiteness, and blue-colored substances did not change their color at all. Reaching for explanations, Dewar and Fleming thought the changes in color corresponded to changes in the substances' specific absorption of light, but they could not be certain. Rounding a corner of the optical-properties valley, the researchers discovered the presence of something they recognized but had not suspected of existing in these latitudes: phosphorescence. All sorts of materials that in the normal-temperature world did not even give off the faintest of shines began to glow with their own bluish-colored light in the extreme cold — substances such as gelatin, paraffin, celluloid, rubber, ivory, and bone. Sulfuric and hydrochloric acids gleamed brightly. An egg immersed in liquid oxygen and then stimulated by an electric lamp radiated as a globe of blue light. Feathers glowed, too, as did cotton, wool, tortoiseshell, leather, linen, even sponges. Perhaps the ability to become phosphorescent had to do with the internal oxygen content of the material, but the experimenters couldn't prove that either. As with geographical explorers encountering strange flora and fauna in a country never before traversed, Dewar and Fleming, in their forays in the temperature region of liquid oxygen, simply captured the beasts, collected the flowers, and carried the samples back home to await further testing and eventual explanation.

"The prosecution of research at low temperatures approaching the zero of absolute temperature is attended with difficulties and dangers of no ordinary kind," Dewar wrote. There existed "no recorded experience to guide us . . . [in] storing and manipulating exceedingly volatile liquids like liquid oxygen and liquid air," which exploded ordinary glass vessels, caused metals to freeze and shatter, and exceeded the measuring capacities of instruments. His rather exalted language couched the difficult though ordinary problem of figuring out how to store low-temperature liquefied gases, and it reflected Dewar's increasingly heroic view of himself as engaged in a great struggle for knowledge.

The solution to the storage problem came to him in 1892, and it harked back to experiments he had begun twenty years earlier, on making vacuums. "Exhausting" the air between the outer and inner walls of a container, he found that the inner vessel could then readily contain the corrosive and volatile liquefied oxygen and could hold it in quantities large enough for a series of experiments. Perfecting this device took him a year and innumerable tries, during which he determined that coating the inside of the flask with a thin layer of silver or mercury reduced loss by radiation by a factor of 13. Dewar presented the first perfected flask to the Prince of Wales at a public meeting at the Royal Institution. The "cryostatic devices" that Dewar produced for his low-temperature work were avidly sought and adopted by everyone else in the field and became known as "dewars." A "magnificent invention," Kamerlingh Onnes called the dewar, "the most important appliance for operating at extremely low temperatures."

The commercial version came about almost by accident: Dewar was having difficulty obtaining proper glass for his cryostats and commissioned a glass blower in Germany to make some for him; that man put his baby's milk in one of the flasks overnight and found that the milk was still warm in the morning. He took the idea of a "Thermos Flasche" to a manufacturer, and an everyday item was born.

Dewar's reticence to patent his own invention has been attributed by some historians to his mingling with the upper crust of London society, whom he might have believed likely to frown on any attempt at commercialism by a serious scientific researcher. But Lord Kelvin was not above obtaining patents for his work, and no scientist was held in greater respect by the elite. Perhaps Dewar did not realize that something designed for use at −200°C would be useful outside the land of Frigor.

In any event, with dewars in hand, their inventor could almost taste the triumph that lay ahead, the true liquefaction of hydrogen; by 1894, it seemed just inches from his grasp.

Seventeen years after Cailletet and Pictet had announced the first, fractional liquefactions of nitrogen and oxygen, Dewar in 1894 stood at the brink of the great mountain range he and everyone else in the field thought was the most important barrier between them and the cold pole of absolute zero. That mountain was the challenge of liquefying hydrogen in quantity. Dewar was working to remove the last kinks and problems from his materials and machinery, since every speck of impurity in the gas supply or minute defect in the seals of the apparatus, or in its ability to maintain pressure, or in the vacuum insulation of the cryostats, could spoil the experiment and result in failure. Although he was not the sort of man who shared his frustrations, the entire scientific community of Great Britain knew that he was on the verge of his greatest triumph and that it might be only a few months before he reached his goal of 20 cubic centimeters of hydrogen boiling quietly in a vacuum vessel, at a mere two dozen degrees above absolute zero.

Repeatedly during the 1880s and early 1890s, while proceeding with his scientific inquiries in the low-temperature region, Dewar would extract from them effects that delighted lecture audiences. The pressure he put on himself to invent better and more startling demonstrations multiplied as the geographic explorers returned from their quests to give popular lectures at other institutions. The Royal Institution's Friday Nighters were treated to having the amphitheater darkened and watching Dewar rub a cotton-wool sponge soaked in liquid air over a large vacuum vessel containing mercury or iodine vapor; just a touch produced luminous glows in the vessel, or bright flashes of light that enabled the audience to see its shape. A bath in liquid oxygen turned oxides and sulfides bright orange, chrome yellow, or metallic white, or made them lose their color. A multicolored soap film, suspended above a flask of liquid air, froze in the dense gas given off, preserving its sequence of colors. Tracing a line of liquid air on a band of India rubber, Dewar made the rubber alternately contract and expand in response to his drawing. He accompanied such magical demonstrations with erudite patter — the changed colors of the oxides and sulfides revealing "that the specific absorption of many substances undergoes great changes at the temperature of minus one hundred and ninety degrees centigrade" — but it was the visual displays that stayed in the minds of audience members.

As a connoisseur of Dewar's lectures later wrote, the showy demonstrations were of interest to "those in his audience who knew what they were witnessing, whilst the rest of his audience was interested much as it might have been by conjuring tricks." Magic shows were at the height of their popularity in Victorian music halls just then. Though the Friday Nighters loved Dewar's showmanship, the more scientifically learned in the audience did not, in general, approve: the theatrics smacked overmuch of magic and illusion, drew attention to the experimenter instead of to the advance of science, and strongly and adversely altered the expectation of lecture audiences for other scientists' reports of their work. Worse, the grandstanding became intertwined with Dewar's growing sense of the importance of his own position and research, his autocratic behavior, and his unwillingness to put enough effort into maintaining good relations with colleagues among the scientific elite.

When Dewar was awarded the Royal Institution's Rumford Medal in 1894 for his dewars, most English scientists applauded, but a few grumbled under their breath at the unfairness of it — could not the committee have also included Fleming in the citation? In Kraków, Olszewski seethed at the announcement. It seemed to him that Dewar had simply copied the metal apparatus he had perfected in 1889 to 1890. Olszewski based his belief on having published a report of the work in a French-language journal in 1890, a copy of which he had sent to Dewar. A long illness in 1892 had made it difficult for Olszewski to leave his laboratory building, and he had simply moved into it his bed and belongings; since he was wifeless and childless, the change in accommodations isolated him all the more and induced in him a touch too much contemplation of his own successes, failures, slights, and annoyances.

Olszewski became convinced that Dewar had deliberately ignored his work, that "the experiments of Professor Dewar are merely the repetition and confirmation of [my] researches," and that "the first apparatus serving to produce large quantities of the liquefied so-called permanent gases . . . was constructed by me." Olszewski further charged that the Dewar-Fleming work on the magnetic properties of materials at low temperatures was just a repetition and slight extension of that previously done by Clausius, Cailletet, and von Wróblewski. These charges were printed in the February 1895 issue of the English-language journal *Philosophical Magazine,* and they appeared in that venue as one result of the growing rift between Dewar and other leading lights of British science.

The bad blood may have dated back to 1877, when the chemist William Ramsay had applied for the chair at the Royal Institution that was shortly won by Dewar. Ramsay was a fellow Scot and was similarly trained; his later successes, including a Nobel Prize in 1904, attest to his strength as a chemist. It is likely that he was denied the Royal Institution position only because he was ten years younger than Dewar. Intimates of Ramsay recall he took the defeat well enough, and he was also turned down for other chairs before settling in at the University of London — but Dewar appears to have never forgiven Ramsay for the effrontery of applying for a chair that he believed was his almost by divine right.

Then there was Lord Rayleigh, born John William Strutt, who was a neighbor of Dewar's, occupying the upstairs laboratory at the Royal Institution. Early in 1894 Rayleigh began to look into an anomaly in the density of nitrogen; within the year, this examination led Rayleigh and Ramsay to the discovery of a new gas, eventually named argon. The day after their first, brief announcement — really a report on work in progress — was made, and again a few days later, Dewar cast doubt on their work in letters to the London *Times,* claiming that what they had discovered might only be an isotope of nitrogen. These letters bothered Rayleigh — he didn't expect to be publicly sapped by a colleague when the information (and doubts) could as easily have been conveyed privately — but Ramsay ignored them and continued on. Dewar pretended to do so as well, even corresponding with Ramsay about his progress in isolating the as-yet-unnamed gas.

William Travers, then Ramsay's assistant, and a brilliant chemist and tinkerer himself, later wrote of this moment that Dewar misinterpreted his own findings and missed discovering argon, concluding that "if he had been skilled on the chemical side, he could hardly have missed krypton, not to speak of neon and xenon," the other noble (inert) gases Ramsay and Travers would discover in ensuing years. In Travers's view, Dewar's strength was the "engineering" aspect of chemistry, not analysis. More important for the story of research into the low-temperature regions, Travers deprecated Dewar's insistence on "a policy of secrecy" about the exact configuration of his apparatus for liquefying air; for most of the early 1890s, that secrecy kept others from being more active competitors in the race to liquefy hydrogen, permitting Dewar to maintain the lead position.

The Ramsay-Rayleigh research continued through the late summer and fall of 1894, when there appeared in the pages of the *Chemical News* letters signed by a pseudonymous "Suum Cuique," suggesting that Dewar, rather than Ramsay, had first alerted Rayleigh to the writings of the professor who years earlier had originally noted the anomaly in nitrogen. It was thought that Dewar, or someone acting for Dewar, was Suum Cuique, but this was never proved.

However, when it came time for Ramsay and Rayleigh to choose a chemist to liquefy argon, as part of a group of people examining its various qualities, Ramsay chose Olszewski rather than Dewar, and Rayleigh could no longer object to this passing over of his neighbor in favor of a distant collaborator. The choice of Olszewski was not mere spite, according to Travers: Ramsay knew Olszewski had trained under Bunsen, as he himself had done, and he had checked Olszewski's results on other matters, which were good; moreover, Olszewski used a gas thermometer for his readings, while Dewar used indirect measuring instruments that Ramsay considered unreliable.

While Olszewski and Ramsay were in contact about the argon research — which Olszewski successfully accomplished — the Pole told his new collaborator about his own long-simmering annoyance at Dewar. In late 1894 Ramsay arranged for Olszewski to publish two articles, one a "Claim for Priority" in *Nature* in January 1895, and a second in the February 1895 issue of the *Philosophical Magazine,* and to announce a forthcoming English-language publication of Olszewski's collected research articles.

Dewar struck back immediately, and with great force, in the next issue of the *Philosophical Magazine:*

> It is usually assumed that if a scientific man has a grievance on some question of priority, he speaks out boldly at or about the time when his discovery is being appropriated. . . . Professor Olszewski prefers to nurse his complaints for four years and then to bring them out simultaneously in two English scientific journals. The result, I am afraid, will be grievously disappointing. . . . We want in this country a reprint of the splendid papers of the late Professor Wroblewski. Until this is done it will be impossible for the scientific public to decide on many of Professor Olszewski's claims for priority.

Dewar then went on to show that one part of Olszewski's apparatus had been taken from an 1878 design by Pictet and that another part had been borrowed from Dewar's own 1886 device, a virtual blueprint of which he had attached to his 1886 article on meteorites. Since that article had been on a seemingly unrelated subject, Olszewski might be excused for not having spotted it; in fact, Onnes had also missed the article — he later wrote that he had overlooked it because no report of the paper showed up in the *Beiblätter* journal of abstracts. But Dewar would not forgive Olszewski for missing his meteorite paper. He also quoted Olszewski against himself, citing other articles in which the Pole had described using glass rather than metal containers up through 1890, and an instance where Olszewski had cited the results of an 1892 Dewar and Liveing article in one of his own reports. Dewar concluded that Olszewski's claims for priority were "fantastic and unfounded."

At this time Dewar also refused to permit Pictet to visit his low-temperature laboratory, in order, he later wrote, "to prevent any further recriminations," and he decided to initiate a correspondence with Heike Kamerlingh Onnes.

In a restrained but determined handwritten note to the Dutch professor in 1895, Dewar wrote that there were only three people in the world who could "know the worries . . . of low temperature research and who can appreciate [that] such work requires a long apprenticeship of a very trying and disheartening kind." He conveyed his upset at Olszewski's articles, which gave to the public the ludicrous idea that liquefaction of gases was easy; Onnes, of all men, would know this was absurd. "The fact is I never learnt anything in the way of manipulation of liquid gases from Prof. Olszewski," Dewar charged. Trying to achieve common ground with Onnes, Dewar confided that his two professorships and their attendant details were getting in the way of his liquefaction research: "If I had nothing else to do but low temperature work I like you might get on faster." He pledged that from here on in, he would do nothing that did not add "lustre to the dignity of science."

Hardly. Not content to leave well enough alone, in an article printed a month after the first salvo at Olszewski, Dewar digressed from the subject at hand to slam the Pole's English sponsor:

One can only wonder at the meagre additions to knowledge that in our time are unhesitatingly brought forward as original, and more especially that scientific men could be got to give them any currency in this country. Such persons should read the late Professor Wroblewski's pamphlet, entitled "Comment l'air a été liquefié" [How the Air Was Liquefied], and make themselves generally acquainted with the work of this most remarkable man, before coming to hasty conclusions on claims of priority brought forward by his some-time colleague.

There could be no doubt about whom Dewar referred to, even if he did not name Ramsay. This unnecessary bashing of a fellow member of the Royal Society, one of the most distinguished scientists of his day, would shortly have repercussions that Dewar could not have imagined, and that would directly influence the forthcoming stages of the race for the ultimate pole of Frigor.

Rare and
Common Gases

B Y 1895 WILLIAM THOMSON HAD BECOME Baron Kelvin of Largs, Great Britain's grand old man of heat and cold, though at seventy he was far from retired. The early experiments he and Joule had conducted together had suffered the usual fate at the hands of time: younger scientists took them for granted and did not reexamine them for clues to further research. The Joule-Thomson effect — the lowering of temperature attendant on the rapid expansion of highly pressurized gases into less pressurized environments through a porous plug — had not attracted much attention from pure-science researchers in the forty years between 1855 and 1895. But in a rare reversal of precedence, researchers with commercial goals in mind paid close attention to the Joule-Thomson effect, as was evident from the near-simultaneous filings of patents for gas-liquefaction processes based on Joule-Thomson expansion in the late spring of 1895 by Carl Linde and William Hampson, which led directly to the first large-scale commercial utilization of the products of the ultracold.

Linde's patent filing was the culmination of nearly twenty years' work, during which his company had sold more than a thousand refrigeration systems and had established its own research laboratory to investigate the commercial potential of newer liquefaction techniques. These pursuits led Linde to combine in a single machine the use of Joule-Thomson expansion, a counterflow principle, and an engine earlier invented by Siemens. Linde aimed at the commercial manufacture of liquefied oxygen and nitrogen, respectively for use in steel making and as agricultural fertilizer.

William Hampson, who patented a similar machine at almost the same moment in time, was a curiosity in Great Britain's scientific circles. Though he had been schooled at Oxford and had trained as a barrister at the Inner Temple, his name never showed up on any lists of lawyers, and his activities before 1895 have remained obscure. They can only be inferred from his later pursuits. He qualified as a medical practitioner and ran the x-ray and electrical departments of St. John's Hospitals in Leicester Square; he also invented electrical devices for muscular stimulation, and he wrote popular science books and an economics tract warning against the use of credit. Hampson came up with his own design for a "regenerative" machine for producing lowered temperatures, based on Joule-Thomson cooling and an adaptation of Pictet's cascade methodology; he was awarded a patent in May 1895, two weeks before Linde. Ralph G. Scurlock, a historian of cryogenics, puts the feat in context by pointing out that "Hampson with his limited facilities was able to invent and develop a compact air liquefier which had a mechanical elegance and simplicity which made Dewar's efforts seem crude and clumsy by comparison." Shortly, Hampson entered into a commercial partnership with Brin's Oxygen Company to produce liquid oxygen.

The efficacy of what came to be called the Linde-Hampson liquefaction process was so evident that pure researchers as well as commercially minded ones immediately sought to adopt the new process, either by purchasing a machine or by using the underlying principles to develop their own versions. Kamerlingh Onnes, for instance, bought a Linde machine as soon as it was available. And in Great Britain, Sir William Ramsay asked Hampson if he could borrow one, because he couldn't obtain any liquid air from James

Dewar. Ramsay and Dewar were at loggerheads again. In December 1895, Dewar told a meeting of the Fellows at the Royal Society about his progress on hydrogen, and Ramsay rose to suggest — once more — that Olszewski had already liquefied the gas. An angry Dewar dared Ramsay to produce proof. At the next meeting, Ramsay had to admit that in the interim he had received a letter from Olszewski, denying having obtained hydrogen liquid from the static form of the gas. Dewar then published his account of the controversy, further humiliating Ramsay.

This occurred at a moment when Ramsay desperately needed liquid air to further his research on rare, inert gases. Just recently he had made a discovery of vast importance to the table of elements and, not incidentally, to the next stages of the exploration of the cold: Ramsay had found helium on Earth. For about a quarter century, helium had been known to exist on the sun, identified through a bright yellow line in a spectrum analysis of the sun's corona. But it had been believed to exist only on the sun. In 1895 Ramsay was working with a sample of pitchblende, a dark rock containing uranium and radium, which was also known to expel argon when heated; he saw in the spectrum analysis of the gases emitted from the pitchblende the same bright yellow line previously seen only in the analysis of the sun, and he concluded that helium was present on Earth in small quantities. There was immediate controversy over helium as well; Dewar thought it might only be an isotope of hydrogen, while Ramsay insisted it was an entirely new element.

To obtain more helium so he could experiment with it and better delineate its properties, Ramsay required liquid air. And since Dewar was now unlikely to give him any, Ramsay looked for other ways to obtain it; he formed alliances with both Hampson and a young researcher at the University of London, Morris Travers, who believed he could devise his own apparatus for producing liquefied gases in quantity. Ramsay bankrolled Travers, but with only £50, a minuscule amount of money compared with what Dewar usually spent on machinery; this budget did not allow Travers the luxury of having parts made, so he borrowed them from other projects — a pipe here, a compressor there — and cobbled them together.

Meanwhile, 'Dewar pushed on by himself, employing Joule-Thomson expansion. He had used it years earlier to produce liquid air, but until 1895 he seemed not to have considered combining it with a "regenerative" process, one that would cool the air further each time it went through the apparatus. When his new apparatus was nearing completion, in early 1896, he described some experiments using it in an article that gave only a nod to Linde and contained a sneering footnote disparaging the mention of Hampson in a report by a French experimenter — even though, as historian Scurlock points out, Dewar's new apparatus owed a significant debt to Hampson's. Dewar also reported that his new machine could only reduce hydrogen gas to much lower temperatures than ever before, but he pledged to soon "overcome" the remaining technical "difficulties" and produce liquid hydrogen.

As Dewar neared his goal, his two main competitors, Olszewski and Onnes, each suffered a misstep. According to Tadeusz Estreicher, then employed by Hampson in London, Olszewski stumbled because his bid to buy a Hampson apparatus failed when the Jagiellonian University would not come up with the funds. Without a new method for lowering temperatures beyond that of liquid oxygen, Olszewski could not produce liquid hydrogen.

Onnes's problem came from a more unexpected source. In mid-1895 the town council of Leiden decided that the use of compressed gases in the low-temperature section of Onnes's laboratory could cause an explosion that would damage the entire building and the surrounding area — and therefore his laboratory ought to be shut and low-temperature research forbidden. Ninety years earlier, during the Napoleonic occupation of the Netherlands, an ammunition ship had exploded in a Leiden canal, destroying about five hundred buildings. The building in which Onnes's low-temperature laboratory was located had been erected on the ruins of the destroyed area of town. "When the town council learned that the laboratory housed considerable quantities of compressed hydrogen, a wildly combustible gas, the historical memory of the ship's explosion drove them into a panic," writes Rudolf de Bruyn Ouboter, a twentieth-century director of the Leiden low-temperature laboratory. As a result of that panic, Onnes's low-temperature work at Leiden was suspended while a commission looked into the matter.

Mortified, Onnes wrote to Dewar in 1896 that he was unable to repeat and verify Dewar's latest "splendid" experiments "for a reason you will be astonished to hear," and he asked Dewar for a favor, pleading, "I do not think you go so far in secrecy that you will not assist a fellow worker." The favor was for Dewar to answer a series of questions about his laboratory in the middle of London, for the edification of the investigating Dutch commission. Dewar responded with good grace, even overstating the case for the relative safety of the equipment by telling Onnes that his own laboratory, where the dangerous gases were held, was directly underneath the auditorium of the Royal Institution; actually, it was two floors below. Dewar also managed not to mention to the commission the explosion in that laboratory that had nearly killed him in 1886. Similar questions that Onnes put to Olszewski were also answered in a reassuring manner.

The key fact presented to the commission was that the explosion of a cylinder of compressed gas would do less damage than the explosion of an amount of gunpowder that it was legal to own and transport. The case went all the way to the Supreme Court of the Netherlands, which decided Onnes could resume his work. By then several years had elapsed, and during the interim Dewar won the hydrogen lap of the race toward absolute zero.

On May 10, 1898, after months of construction, testing, and preliminary trials, Dewar and his two assistants, Robert Lennox and James Heath, cranked up the jumble of pumps, cryostats, tubing, gas-intake valves, and liquid-collection points that filled the basement of the Royal Institution, making the space more resemble the boiler room of a factory than a scientific laboratory. The apparatus was a series of liquefaction machines. The first set, or step, produced chloromethane, which was used to cool the ethylene of the second step, which was itself then used to cool oxygen to the point of liquefaction. That liquid oxygen became the primary coolant for the attempt at liquefying hydrogen. After the hydrogen had spent many cycles in the regenerative process, Dewar and his associates succeeded in cooling the gas to −205°C. Then, while the gas was under a pressure of 180 atmospheres, they released the hydrogen suddenly and continuously into a vacuum vessel that was surrounded by a −200°C atmosphere and through that vacuum vessel into a second encased by a third. Within the space of five minutes, Dewar shortly reported, twenty cubic centimeters of liquid hydrogen were collected. The liquid was clear and colorless, with a readily observed meniscus (curved upper surface). After five minutes, however, "the hydrogen jet froze up, from the accumulation of air in the pipes frozen out from the impure hydrogen," and the experiment was forced to a stop.

Dewar was elated to have liquefied the last of the permanent gases. Wanting to prove to himself that this liquefied hydrogen was colder than any substance yet available, he plunged into the new liquid a tube filled with liquid oxygen; the oxygen froze to a bluish white solid, which could only happen if the temperature of the new liquid was lower than that of the oxygen. Dewar believed for a while that he had also managed to liquefy helium. He claimed that feat in a telegram to Onnes — "Liquefied hydrogen and helium" — but shortly he decided that the additional condensate was from traces of impurities in the hydrogen. Perhaps to discourage Kamerlingh Onnes from returning to the race, Dewar then accentuated the difficulties, writing to him in November 1898, "My troubles I can see are only beginning. It will be a long time before Hydrogen is *on tap*."

Another problem: The platinum-resistance thermometer did not give completely trustworthy readings of the boiling point of hydrogen; upon reaching what Dewar believed to be that point, the platinum thermometer seemed to stop working, and Dewar wondered if it had "arrived at a limiting resistance," below which the changes in resistance had become too small or too difficult to measure. Seeking some positive information from the thermometer's failure, he concluded — as he later wrote — that there was "no longer any reason to believe that at the absolute zero, platinum would become a perfect conductor of electricity," or by extension, that any other pure metal would do so. This contradicted the earlier prediction of his collaborator, Fleming, but Dewar did not try to prove his new contention, or to construct a theory to explain the apparent failure of the resistance thermometer.

Instead, Dewar used liquid hydrogen to accomplish what had previously been impossible. A year earlier, he had teamed with Henri Moissan, the discoverer of fluorine, to liquefy that gas. They could not do it at liquid-oxygen temperatures, but now, using liquid hydrogen, they were able to solidify fluorine. Liquid hydrogen also allowed them to determine that fluorine, unlike many other chemicals, remained reactive at very low temperatures: when they directly mixed solid fluorine and liquid hydrogen, the mixture violently exploded.

Since Dewar was still uncertain as to the precise temperature of liquid hydrogen, he searched for a thermometer to measure it and found one of an entirely new type, containing gaseous hydrogen under pressure. Using it, Dewar estimated the boiling point of hydrogen at 20 to 22 K, or −250°C. "With hydrogen as a cooling agent," he confidently predicted, "we shall get to from 13 to 15 of the zero of absolute temperature, and . . . open up an entirely new field of scientific inquiry."

A few days after Dewar had made his preliminary communication about the liquefaction of hydrogen in quantity, William Hampson gave a lecture on the "self-intensive refrigeration of gases," which Dewar attended. Dewar challenged the upstart's claim that in addition to liquefying air and oxygen, he had liquefied hydrogen. Hampson reiterated his claim in the next issue of *Nature* and added that several years earlier, he had gone to the Royal Institution and briefed Dewar's assistant, Robert Lennox, about the machine he was perfecting; Hampson implied that Lennox had then told Dewar, and that Dewar used the information to construct his hydrogen-liquefaction apparatus. Dewar thundered in the following issue that he would have liquefied hydrogen in the same way at the same time even if Hampson had never been born. Three additional letter thrusts from Hampson and three more ripostes from Dewar followed, in subsequent issues. The conflicting claims were never resolved, and today one can only conclude that both men had right on their side: Dewar clearly had some knowledge of Hampson's work, perhaps from Hampson's patent application if not from his visit to Lennox; but equally, Dewar's search for a way to intensify the cold had led him to the same principles of Joule-Thomson expansion and the regenerative cycle that Hampson — and Linde — adopted. In later years, Dewar would go out of his way to avoid mention of Hampson in his occasional histories of low-temperature research, just as he also attempted to erase Olszewski's name from those histories. These unseemly overreactions appear to have been mandated by Dewar's obsession with his low-temperature goal and his egotism about the worth of his achievements.

A painting of Dewar lecturing, presumably on the hundredth anniversary of the Royal Institution in 1899, still hangs on the institution's walls today. The solemn stocky man, in his frock coat and pointed beard, is behind a table full of flasks, burners, and small apparatus, and underneath a screen that features projected images of the equipment. He holds a vacuum flask at arm's length before him, while Heath and Lennox behind him ready other materials for the demonstrations. Rapt attention is focused on Dewar by an elegantly attired audience composed of such distinguished visitors as Marconi, and the artistic, scientific, and political elites of the British Isles, including Kelvin, Stokes, Lord Rayleigh, and Rayleigh's brother, Prime Minister Arthur Balfour.

The lecture was about liquefying hydrogen, and in it Dewar

crowed that "Faraday's expressed faith in the potentialities of experimental inquiry in 1852 has been justified forty-six years afterwards by the production of liquid hydrogen in the very laboratory in which his epoch-making researches were executed." For this first public demonstration of liquid hydrogen, Dewar the showman pulled out all the stops. He dipped a piece of metal in liquid hydrogen, then removed it; air instantly condensed around the metal, forming a solid coating that then melted to a liquid, which he collected in a cup. Into that cup he inserted a red-hot splinter of wood, and the wood ignited as oxygen evaporated from the liquid air. Using the liquid hydrogen, he produced phosphorescence in all sorts of substances, decreased electrical resistivity in metals, sent ordinary thermometers shooting down until they failed, turned liquid oxygen blue, showed that liquid hydrogen was fourteen times less dense than water, and used liquid hydrogen to magnetize cotton wool.

Dewar closed by expressing his gratitude to Robert Lennox in the most fulsome words he ever used publicly for an employee, avowing that "but for his engineering skill, manipulative ability and loyal perseverance, the present successful issue might have been indefinitely delayed," and by giving thanks to the members for supporting his work, coupled with a warning that future research in the extreme cold would be immensely difficult and very expensive.

The notable British scientists from the era of Bacon, Boyle, and Newton in the seventeenth century down through that of Davy, Faraday, and Kelvin in the nineteenth had for the most part disdained commercial endeavor as beneath the dignity and beyond the purview of the basic researcher. But the heyday of that notion had passed, along with the time when the lone scientist in his personal laboratory could make significant discoveries. Across the English Channel, Dewar's rival Kamerlingh Onnes was in the process of establishing the first "big science" endeavors in the low-temperature physics laboratory at Leiden, with its ancillary school for instrument makers and other assistants; Onnes was also moving in the direction of encouraging interchange between his laboratory and the research and development facilities of manufacturers who wished to exploit the commercial potential of the cold.

While James Dewar's battle with Hampson did indeed knock that particular amateur out of the scientific race toward absolute zero, Dewar accomplished little by doing so, for on the pure-research front, Hampson continued to assist Ramsay, Rayleigh, and their aide Morris Travers in isolating the remainder of the noble gases. And on the commercial front, Hampson also forged ahead.

The first several steps in the liquefaction cascades could now be commercially replicated. Combined with the Linde and Hampson patented processes of 1895, these cascade steps made it possible to use cold to manufacture liquid air and separate from it liquid oxygen and liquid nitrogen, also producing liquid traces of many other elements and compounds. In short, while the scientists of the pure-research laboratories pursued rare gases such as argon and helium, the commercial scientists and technologists tried as hard as they could to make liquefied gases the stuff of everyday life.

The first major technology of the cold, for refrigeration, was spreading, especially in America. Back in the 1840s, John Gorrie had wanted to use refrigeration of air to "counteract the evils of high temperature, and improve the condition of our cities," through the use of central refrigeration plants that would pipe cool air to homes and businesses. Centuries earlier, Drebbel had proposed something similar for heat distribution in London. By 1889, cooled air piped from a central station was available in New York, Boston, Los Angeles, Kansas City, and St. Louis; in the latter city, the proprietor of the Ice Palace restaurant had such frosty air piped in that in addition to cooling his customers he was able to use it to spell out his name in ice on his window; he reinforced the cold's effect by wall murals of scenes from polar expeditions. The exceptionally mild winter of 1890, during which relatively little natural ice formed, spurred the further progress of artificial icemaking by intensifying demand for mechanical refrigeration among manufacturers of various products. But all those who wanted better refrigeration had to await the development of more efficient and less dangerous coolants.

After 1895, the main manufacturing firms for liquefied gases, used as coolants or for other purposes, were Linde's, based in Germany, and the British Oxygen Company, successor to Brin's, which had made an arrangement with Hampson. Some rivalry developed when Linde's concern opened a British branch, competing directly for a time with the British Oxygen Company.

As for the United States, there was a flurry of excitement in June 1897 when a *New York Times Magazine* article opened with this memorable sentence: "Mama wants two quarts of your best liquid air, and she says that the last you sent had too much carbonic acid gas." The article referred to American engineer Charles E. Tripler and his recently announced steam-driven machine for the liquefaction of air. It was an opportune moment to begin such an enterprise, because contemporary internal-combustion engines were considered unreliable, and therefore unsuitable for the new horseless carriages — but the already proven technology of air-expansion (compressed-air) engines could provide the horsepower. Tripler's promise of producing large quantities of liquid air for such engines in carriages, ships, and other modes of transportation attracted Wall Street investors. In short order, with the help of some stock salesmen, Tripler raised $10 million for his public company. The engineer proved a good promoter, able to help his cause by giving lectures and interviews. Something of a visionary, he predicted additional uses for his liquid air: in refrigeration; in explosives, since with powdered charcoal, it could produce quite a bang; and in medicine, where, Tripler said, it had already been tested as an antiseptic in surgery and was believed to hold promise as a cure for cancer.

Tripler was viewed as perhaps too visionary, and many people vigorously debunked him for an over-the-top boast in *McClure's*, in which he had told a well-respected writer that liquid air was "a new substance that promises to do the work of coal and ice and gunpowder at next to no cost." His machine evidently did work; he sent some of his product to a University of Pennsylvania chemist, who verified that it was indeed liquid air. Apparently, though, Tripler knew so little about chemistry and physics that he dared to assert he had fed 3 quarts of liquid air into his machine, and because of cold's ability to produce additional cold through evaporation, he had been able to obtain 10 quarts of liquid air from the energy provided by 3 quarts.

The ability to use a liquid gas's own coldness to cool it further, toward a solid, was something that James Dewar liked to demonstrate in his lectures. But Tripler's use of this self-intensification of cold emboldened him to utter what was virtually a new claim of perpetual motion. While legitimate scientists might have continued to ignore Tripler and let him go on to produce liquefied gases without comment, they could not let stand the idea of someone selling what was, in effect, the snake oil of perpetual motion. They proceeded to jump all over him. *Scientific American*, which had earlier reported Tripler's feat and the testing of his products by the University of Pennsylvania, now printed the comments of the president of the Stevens Institute of Technology, who pointed out that the second law of thermodynamics made three-for-one production impossible. In 1899, with the controversy still going on, Harvard's senior physicist commented that Tripler's use of one to make three "will succeed only when water is found to run up-hill." Linde, on a lecture tour of the United States, also debunked the claim.

The stock of Tripler's firm collapsed and shares became worthless; shortly, investigators found that most of the $10 million had gone into the pockets of the promoters and not into Tripler's manufacturing process. In reaction to this debacle, American businesses refused to have anything to do with the commercial use of liquefied air for some years thereafter.

In France, the early history of liquefied air almost went the same route, but it did not, because Georges Claude was a better scientist than Tripler. The son of an engineer who was also the assistant manager of an ice cream company, Claude had been involved with the cold all his life. After his university training, he became fascinated with liquefied gases in the early 1890s, when he worked as an electrical engineer and chief of a laboratory at Les Halles, Paris's marketplace. At that time, liquefied acetylene was used in welding torches. Claude envisioned replacing the acetylene with a torch of carbon burning in pure oxygen, which he believed would shortly be separated out of air; at that moment, when liquid oxygen had been produced mainly in basic research laboratories, this was a fairly farsighted idea for anyone not part of the small circle of low-temperature researchers.

In 1895 Claude learned about Linde's separation machine. The existence of such a machine thrilled him, and he became certain he was destined to do nothing less than establish the entire industry for the manufacture of liquefied gases. To accomplish that, he would have to operate at the forefront of science, he wrote, which would require great intellectual energy and "brutal perseverance, the daughter of obstinacy and the sister of stubbornness and bad temper." Perseverance meant putting up his own money and convincing friends to loan him theirs, for a total of 50,000 francs, with which he formed a company, not to manufacture liquefied gases right away but first to test a variety of production equipment. As Claude later wrote in a memoir, he came to the key insight by questioning what Linde had done: "Why does he expand his air through a single tap? If this air was made to push a piston, it would produce more work and consequently more cold." He concluded that Linde had avoided the piston method because he was "afraid — quite rightly — that the normal lubricants would freeze [at the very low temperatures] and block his machine." Claude searched for a better lubricant. On the evening of May 25, 1902, with the syndicate about to expire the next day, incurring the loss of all of the 50,000 francs, he made a final improvement to the lubricant and machine, and the whole process worked beautifully, using only 20 atmospheres of pressure to make liquid air, rather than the 200 atmospheres of pressure used by Linde's apparatus.

Claude's process was not merely an incremental technological advance, historian Ralph Scurlock contends; it was a "mechanical revolution so large as to constitute a second technological breakthrough" in the field that would soon become known as cryogenics. The magnitude of the breakthrough was obvious enough for Claude to form a new company within two months, with a capital of 100,000 francs, christened L'Air Liquide, Société Anonyme pour L'Etudie L'Exploitation des Procédés Georges Claude.

Its products, and those of the Linde and British Oxygen groups, were immediately in demand. Steel producers saw that the use of liquid oxygen could upgrade the old air-blast furnaces, helping them produce steel of greater purity and tensile strength, at lower cost. Liquid nitrogen quickly became essential to the production of calcium nitrate fertilizer, ammonia, and saltpeter — the latter product used for explosives, as Tripler had predicted.

Within a short time of their establishment, the companies founded by Linde and Claude found that the appetite for their products was large, and continuing to grow. Along with artificial refrigeration and artificial manufacture of ice — industries that were also continuing to grow by leaps and bounds — the new liquefaction manufacturers made it likely that the twentieth century would be the first ever in the history of the world to be characterized by extensive commercial exploitation of the cold.

The Fifth Step

I N THE CHRONICLES OF many geographic explorations, there comes a moment when a far prize moves into clear view, and all pretense at examining the intervening landscape drops away, in favor of throwing every available resource into the thrust to reach the goal. In the exploration of the cold, that moment arrived when hydrogen was liquefied in quantity and the existence of helium on Earth had been revealed. On learning about the availability of liquid hydrogen, Kamerlingh Onnes later wrote, "I resolved to make reaching the end of the road my purpose immediately." The goal that became so tangible was the possible liquefaction of helium, which all the competitors in the race now believed could be made to happen at less than 10 degrees above absolute zero.

In 1898 Dewar solidified hydrogen by an ingenious method. Long ago, he had begun research on the absorptive power of charcoal; in more recent years, he had again taken up that idea and had used charcoal to absorb hydrogen-gas molecules and separate them from the liquid. This intensified the cold, forcing the hydrogen to solidify. Then, applying pressure to solid hydrogen, Dewar reached down precisely as far as he had predicted, to within 13 degrees of absolute zero. "But there or thereabouts," he wrote, "our progress is barred."

Preliminary analysis showed that helium resisted liquefaction even at −260°C. Because scientists had already descended hundreds of degrees from normal temperatures, the remaining 13 degrees might seem to a layman only a slight distance to travel. "But to win one degree low down the scale," Dewar wrote, "is quite a different matter from doing so at higher temperatures; in fact, to annihilate these few remaining degrees would be a far greater achievement than any so far accomplished in low-temperature research."

In a presidential address to the BAAS, Dewar painted the liquefaction of helium as the next great goal of science, the last stop on the way to the Ultima Thule, absolute zero. He also predicted the characteristics of liquid helium: it would have a boiling point of about 5 K; it would be twice as dense as liquid hydrogen and seventeen times as dense as gaseous helium; it would have a negligible surface tension; and it would be difficult to see.

Dewar was able to imagine the mysterious, pristine, frigid glory of liquid helium, but to liquefy it he required gaseous helium in quantity, and he didn't have enough. Supply also became of prime importance for Onnes, Olszewski, and others still in the race for the cold pole. In Great Britain, the largest concentration of helium had been found in the mineral springs at Bath. By a combination of circumstances and connections, Ramsay for a while controlled access to this source of helium, which was a problem for Dewar. But Dewar had the only apparatus that could provide the liquid hydrogen Ramsay needed to do further research on the inert gases. Several scientists, Dewar among them, had postulated the existence of another inert gas with a molecular weight between that of argon and helium, and Ramsay was working hard to find it. Ramsay and Dewar could have made a nice trade, helium-shale for liquid hydrogen, but their personalities wouldn't permit it, and they were at a standoff.

In 1901 the impasse was broken when Morris Travers of Ramsay's laboratory successfully constructed a hydrogen liquefier; in an article about it, Travers acknowledged help from Hampson, and he also pointed out that the apparatus had cost a mere £35, which his readers understood to be a slap at Dewar, whose apparatus had cost thousands of pounds, and who had fretted in public about the high cost of conducting low-temperature research. Actually, Travers had cannibalized or borrowed equipment whose value was the equivalent of thousands of pounds; liquefying hydrogen could not have been done at that time with available techniques and £35 worth of materials.

During the next few years, using liquid hydrogen as their investigative tool, Ramsay, Rayleigh, and their group discovered xenon, neon, and krypton, completing the array of "rare" gases that would win the two major researchers Nobel Prizes in chemistry and physics in 1904. (Dewar chafed at that, among other reasons because neon was positioned on the periodic table precisely where he had predicted it would be.) In the midst of this research, Ramsay, Travers, and Hampson attended a party given by one of Dewar's friends, even though they did not want to, to prevent their rival from guessing by their absence that they were on the verge of a discovery. Ramsay legitimately feared that Dewar would find new rare gases before he did, based on Dewar's comments when Ramsay or Rayleigh would cautiously speak about their progress at scientific meetings. Travers would later charge that Dewar had not been an "imaginative" enough chemist to have found the rare gases. More likely, Dewar was too intent on liquefying helium just then to pay attention to any other goal. Dewar had a reciprocal, legitimate fear of the Ramsay-Travers group, because Travers's liquefier put them in contention in the race for absolute zero, along with Olszewski and Onnes, who soon produced liquefied hydrogen and aimed at liquefying helium.

During the next several years, the lead in this race changed hands several times among these groups, with each edging closer, down to within 10 degrees of absolute zero. Later on, in 1906, Travers designed an apparatus to liquefy helium, but he subsequently lamented, "Just when all this was ready I had to leave for India, and the experiment was never carried out."

Dewar sensed he had the most to fear from the laboratory at Leiden. Since Heike Kamerlingh Onnes had begun the experimental physics lab there in 1882, it had always been a step behind in the race, but Onnes had never permitted the distance between it and the leader of the moment to widen. A quarter of a century had passed, and although Onnes's hairline had receded to a thin fringe at the back of his pate, and his mustache had come to resemble that of a walrus, he had developed into a pillar of the scientific establishment, a man cherished for his strict regard for measurement, and for taking things one step at a time. Virtually alone among experimental physicists of his day, he had set out on a deliberate program of what would today be called "big science," the establishment of substantial laboratory sections and cadres devoted to thermodynamics, electricity, magnetics, and optics. "He ruled over the minds of his assistants as the wind urges on the clouds," recalled physicist Pieter Zeeman, who trained under Onnes. "He could achieve miracles with a flattering remark or witty (sometimes biting) irony. Even those who were above him on the hierarchical ladder fell for his charm . . . and at the last moment a decision beneficial to Onnes could sometimes be achieved."

While overseeing and developing the other sections of the laboratory, Onnes himself relentlessly pursued the liquefaction of gases. His ego did not keep him from adopting whatever innovation was made by another competitor, then improving it greatly — for instance, when he finally set up his own continuous-process hydrogen liquefier, it produced 4 liters per hour, a quantity vastly outstripping what Dewar, Olszewski, or Travers could make from their machinery.

Similarly, Onnes had nurtured an additional strength: in 1901, adjacent to his laboratory building, he had begun the Society for the Promotion of the Training of Instrument-Makers, a workshop and school that supplemented the one he had previously established for machinists and electro-technicians, and he had set these men to work improving his facilities and machinery. Also, the low-temperature laboratory at Leiden was now beyond worries about cost; Onnes enjoyed the confidence of Queen Wilhelmina and the fullest of government support for his work. His equipment was so large and of such good quality, and his attendant assistants were so well trained, that physicists from all over the world vied to spend a sabbatical or a summer at Leiden. Even the young Albert Einstein applied for a position; he never received an answer to his letter to Onnes, and he went on to other pursuits. Later, the two men would develop a friendship.

As Onnes aged, the tubercular weakness in his lungs, evident since childhood, was exacerbated — perhaps by the noxious gases of the laboratory — and he struggled to be well enough to work; his wife's assistance in maintaining his health was widely admired. Onnes lived in Ter Wetering, a beautiful home by the Old Rhine river with a lovely view, but for years was not well-to-do. As he would recall in a letter to van der Waals, he endured decades of poverty, "care-ridden years [with] appalling difficulties." The theorist comforted Onnes with the saw that "time sifts admirably well that which has real value from the other things," suggesting that vindication of the greatness of Onnes's work, as well as of van der Waals's own theories, would eventually arrive.

Some in the Netherlands chided Onnes for being too patient, for taking too many measurements before thrusting ahead in liquefaction. He ignored the cavils, confident of eventual applause, and continued to build his organization, his apparatus, his communications, his network of supporters in other European laboratories, and his storehouse of knowledge.

Dewar's recognition of Onnes's strengths may have been one reason he was always courteous to Onnes, even during the years he was savaging Olszewski and Ramsay. Dewar pleased Onnes by repeatedly referring to van der Waals as a genius underappreciated in countries beyond the Netherlands. For his part, Onnes went out of his way in his publications to pay tribute to Dewar's accomplishments.

The two men were perhaps most alike in being autocratic, old-style masters of laboratories. Dewar refused to give much credit to his assistants, even though both Robert Lennox and James Heath each lost an eye during their long service with him, and he would not permit his assistants to become familiar, rebuking another associate who dared to call him "Sir James" after Dewar had been knighted. Similarly, in later years a top-hatted Onnes would drive with his horse and wagon to pay Sunday social calls on his technicians. During one such visit, an assistant asked for a raise; Onnes refused, saying "Haven't you just bought two nice bicycles?" When an idea occurred to Onnes at home, he would ring a bell on the side of his house, summoning his chief assistant Gerrit Flim, who lived on the opposite bank of the Old Rhine, and who would have to drop what he was doing and row across to discuss the idea with his boss. Although Onnes treated full colleagues with complete deference, there are stories that he was less respectful of doctoral or postdoctoral students and that he seldom shared credit with them on articles; he also required students to donate part of their stipends for the upkeep of the laboratory. One subsequently recalled their master's rules: they should work all day in the lab, taking notes in a little pocket notebook, and in the evening should write up their notes into a neat report. A student could have "a day off for love; I mean when you were engaged. But you could not get a day off really to listen to a theoretical lecture because [Onnes believed that] experimental physics requires the whole man." Other stories, however, paint Kamerlingh Onnes as the perfect colleague, generous of spirit, and the catalog of his publications shows that as he aged, he became progressively more willing to share credit for the work done under his direction.

In 1902, the second year Nobel Prizes were given out, the physics award went to H. A. Lorentz, a friend and colleague at Leiden, and to Pieter Zeeman, for whom Onnes had served as thesis advisor; the award was for their discoveries regarding the theoretical basis for the splitting of the spectral lines of sodium in a magnetic field. Onnes had as yet made no achievements of similar import, though it was widely believed that he would soon do so. In honor of Zeeman and Lorentz, Onnes commissioned his nephew Harm, a noted artist, to make a stained-glass window memorializing their work.

Dewar and Onnes differed in one critically important area: Dewar's continual stream of published papers made very little use of others' theories — including that of van der Waals — and Dewar wrote no theoretical papers of his own. Onnes was continually guided by theory, and he spent much of his time making calculations from theory to predict isotherms that would reveal the critical temperatures of gases to be liquefied, then experimenting to verify the predictions. An exchange of letters in 1904 highlights this difference. When Onnes congratulated his rival on becoming Sir James, Dewar chided him for going overboard, because he, Dewar, was neither a baronet nor a knight, then went on to declare in the same emotional tone, "Exact physical measurements and pioneering work do not go well together. Such refined matters I must leave you to settle." To which Onnes replied, "The determination of the isotherms is the rational way to get data for calculating the critical points . . . and exact determination of isotherms is just in my line of accurate measuring work."

In preparation for his assault on absolute zero, Onnes pushed hard to determine the isotherms of hydrogen and helium. Later he would look back and view these efforts as the most crucial to his accomplishments, writing, "Before the determination of the isotherms had been performed I had held a perfectly and different opinion [of the possibility of liquefying helium] in consequence of the failure of Olszewski's and Dewar's attempts" to do so. After determining the isotherms, however, he became convinced he was on the right path.

On the practical front, Onnes — like Dewar — was hampered by the lack of an adequate supply of helium gas. The two had traded letters about helium for some time, and an Onnes assistant, acting as an emissary from Leiden, had even visited Dewar's laboratory. Each scientist complained to the other of bouts of ill health that kept them from their research, and of the "fairly risky business," as Dewar put it, of working with helium. "I have already lost 1 cylinder of helium by the breaking of vacuum vessels during the course of its circulation at liquid air temperatures and I dread any repetition of the disaster," Dewar wrote. Worried he would not be physically well enough to complete the work, he penned a poignant wish to Onnes to possess again "the gift of youth so that I might begin my scientific career after a training in the great Dutch School of Science."

Perhaps the comradely and confessional tone of Dewar's letters encouraged Onnes in mid-1905 to take the extraordinary step of proposing to Dewar that they join forces. Dewar had built a plant to extract helium from the Bath Springs sands; Onnes begged to share that material with him, and he implied that they might jointly work through the "determination of the isotherms of helium at low temperatures as well as the magnetic dispersion of the plane of polarization."

"We both want the same material in quantity from the same place, at the same time, and the supply is not sufficient to great demands," Dewar wrote back, in anguished but firm handwriting. He added, "It is a mistake to suppose the Bath supply is so great. *I have not been able so far to accumulate sufficient for my liquefaction experiments.* If I could make some progress with my own work the time might come when I could give you a helping hand which would give me great pleasure." Then, perhaps remorseful that he had to be so tough with Onnes, Dewar confided that things in London were "in a sad way," that his illness prevented him from doing much work, and that he hoped to get away for a rest. The regretful tone continued in a note a month later, in which he apologized for having not cited Onnes's work in an article.

Dewar's refusal to share the Bath helium probably had more to do with his ego than with the difficulty of obtaining helium. He was not averse to collaboration, having toiled for many years with various scientists on jointly signed work. But he had been laboring by himself for nearly thirty years on reaching absolute zero, and he was not about to change his solo style when the competitors were on the last lap of the race.

Dewar's decision not to share the Bath Springs helium with Onnes may have prevented a competitor from using that source, but he had no joy of it, because when he transported the Bath Springs helium to the Royal Institution, he found it contaminated with other gases and needing further purification, which slowed his progress. Just when Dewar's disappointment was becoming obvious, Onnes found his own supply of helium.

Through the auspices of his brother Onno, who was in charge of the Dutch government's Office of Commercial Intelligence, in 1905 Onnes arranged to import large quantities of monazite sand from gravel pits in North Carolina, sand known to contain significant amounts of rare-earth metals and helium. After that, as Onnes later wrote, "the preparation of pure helium in large quantities became chiefly a matter of perseverance and care." That was an understatement, since the process of extracting helium gas from the sand was complex, involving exploding the sand with oxygen, using liquid hydrogen to freeze out certain gases, and compressing the helium gas until it was absorbed by a charcoal filter, from which it could be recovered. Four chemists, working continuously over the next three years, were able to produce barely enough helium for experimental purposes.

During that period, Onnes remained uncertain that his attempt to liquefy helium would succeed, because scientists believed that at very low temperatures the Joule-Thomson effect might not further lower the temperature of a gas but might actually raise it. More perplexing was the unknown temperature at which the gas would become a liquid. All three laboratories had used a test devised by Dewar to measure how much helium a charcoal filter would absorb at the temperature of liquid hydrogen, but the results differed. Olszewski guessed that the critical temperature of helium might be less than 2 degrees above absolute zero, Dewar now thought it would be closer to 8 degrees above, and Onnes's own best estimate was 5 to 6 degrees.

Onnes recognized that the differences between 8, 5, and 2 degrees were highly significant, and he wrote that the results of the various laboratories' charcoal tests not only "left room for doubt" in his mind; far worse, they begat "ample room for fear that helium should deviate from the law of corresponding states." For a quarter century, Onnes had been conducting low-temperature research to experimentally verify the theory of corresponding states articulated by van der Waals. If the critical temperature of helium was really significantly lower or higher than the 5 K predicted by van der Waals's law, then that law of corresponding states would not apply to all elements, which would mean it was not universal, and his friend's theory might have to be junked. Moreover, in practical terms, a 2 K critical temperature would also mean that even the most advanced state-of-the-art apparatus would not be able to force the gas to become a liquid. "So it remained a very exciting question what the critical temperature of helium would be. And in every direction . . . we were confronted by great difficulties."

It was at this moment that Kamerlingh Onnes's determination of the isotherms rescued him from paralyzing doubt, because the mathematical calculations showed that the temperature at which helium ought to liquefy was between 4 and 5 degrees above absolute zero — not, as he had feared, significantly higher or lower than that. Dewar now began to agree with him that the critical temperature was more likely to be nearer 5 K than 8 K.

Onnes pored over the details of the work of Hampson and Olszewski, trying to figure out whether his adaptations of their machinery would do the job. Once, in March 1908, he thought that he had accomplished the task, but in an odd way — that he had compressed helium into becoming a solid, without its ever having stopped at the liquid state. He dashed off a telegram to Dewar to this effect, and received a congratulatory telegram in return, before having to retract his claim because the solid he had produced was composed not of helium but of impurities in the gas.

Before the retraction reached the Royal Institution, Dewar had written a letter to the London *Times* noting the feat. The letter was published, and then Dewar had to apologize for it to Onnes: "I felt a duty to inform the world . . . that you had succeeded where I failed. Considering the enormous difficulty of such experiments we can all be mislead." In this letter to Onnes of April 15, 1908, Dewar all but conceded that Onnes would shortly win the race. He expressed admiration for Onnes's ability to admit mistakes, to identify the problem that caused them, and to move ahead. As for himself, he wrote, his health was improving, though only slowly; more to the point, "the Royal Institution has no money to prosecute such many extensive experiments . . . [and] has no endowment to draw on." What Dewar did not inform Onnes directly about was a near catastrophe in his lab: during an attempt at lowering the temperature of helium, the impurities in the gas froze and clogged a capillary tube; in a too-quick reaction, one of Dewar's assistants turned a vent the wrong way, and many months' supply of hard-won helium gas escaped into the upper atmosphere. This, Dewar recognized, was fatal to his chances to win the race.

Onnes, chastened himself, wrote Dewar a long letter detailing the reasons for his mistaken claim of liquefaction, and in an article he recorded that during the preliminary trials, he had found that "the utmost was demanded of the necessary vacuum glasses," and he worried that "the bursting of the vacuum glasses during the experiment would not only be a most unpleasant incident, but might at the same time annihilate the work of many months." To Dewar, Onnes explained that when he had produced what he thought was a liquid condensate in the midst of a cloud of helium gas, "the tube broke and so I could not have more certainty about the nature of the cloud."

On July 9, 1908, preliminary operations toward the liquefaction of helium began at Leiden, with the cranking up of the first three steps of the cascade, which used chloromethane to liquefy ethylene, the ethylene to liquefy oxygen, and the oxygen to liquefy air. Then the liquid air was used in the fourth step, to make liquid hydrogen.

July 10, 1908, began early for the Low-Temperature laboratory at Leiden; at 5:45 in the morning, preparations started for the fifth step, the liquefaction of helium. As Onnes would shortly write in *Communication No. 108*, everything was in place for this descent: the prepared bottles of liquid hydrogen, the supply of helium purified from the monazite sands and held "ready for use in silvered vacuum glasses," a gas thermometer based on helium gas kept at low pressure, the heat exchangers, the tubing and stopcock apparatus for using Joule-Thomson expansion, the adapted Cailletet compressor with a mercury piston for pressuring the gas — seven years had been required to put it into working condition — and the vacuum cryostats, reinforced glass vessels into which the results of the experiment would flow. The apparatus consisted of both massive iron components and delicate pipettes, boilers and cryostats, sealed bolts and fan belts, slabs of metal and thin wires. Onnes, his chief assistant Gerrit Flim, some colleagues, and several "blue boys" (instrument-maker students) had rehearsed the procedure many times.

Some among the onlookers knew that this laboratory was just steps away from the spot where in the eighteenth century the great chemist Boerhaave had taught his students about the near boundaries of the country of the cold, using as a text Boyle's *Experiments Touching the Cold*. Now the explorers were nearing the epicenter of that country.

It took Onnes's team until 1:30 in the afternoon to make certain that the helium in the apparatus had been entirely cleared of the last traces of air, "by conduction over charcoal in liquid air" through a side conduit. Every stray nook and cranny of the apparatus was filled with liquid hydrogen to protect from the unwanted influx of heat, and to purge any atmospheric air left in the apparatus, as that would have solidified at the low temperatures needed for the lique-faction of helium, producing a snow that could have clouded the glass and made observation of the liquid helium impossible.

At 3:00 in the afternoon, work was so intense that when Onnes's wife, Elisabeth, showed up with sandwiches for lunch, he could not stop to eat them, and Elisabeth fed them to him, bit by bit, as he gave instructions, turned dials, and watched gauges.

They were using a thermometer based on helium — employing helium gas at low pressure to measure the temperature of helium gas approaching liquefaction, a technique dependent on the pres-sure-volume equation in a way that might have delighted Robert Boyle. Since pressure multiplied by volume is proportional to tem-perature, and since the volume of helium gas in the thermometer was constant, measuring the pressure revealed the temperature.

At 4:20 in the afternoon, the apparatus and the helium gas in the canister were both at the proper temperature, −180°C. A gauge was turned to let helium gas into the apparatus, and the protecting glass was filled with liquid hydrogen. Since the experimenters could not see into the interior of the apparatus, the thermometer alone would tell them what was happening inside. At first, Onnes wrote, "the fall of the helium thermometer which indicated the temperature under the expansion cock, was so insignificant, that we feared it had got defect [sic]. . . . After a long time, however, the at first insignificant fall began to be appreciable, and then to accelerate."

They added more liquid hydrogen, and increased the pressure on the helium; by 6:35 in the evening, the temperature for the first time fell below that of liquid hydrogen. Combinations of more and less pressure, and varying expansion volumes, were tried; the ther-mometer once dipped as low as 6 K, then wavered upward.

By this time, word had gotten out to scientific colleagues at the university that the critical moment had come in the Low-Temperature laboratory, and people began to drift in to watch. Among them was Professor Franciscus Schreinmakers. Onnes was calm, but he could not refrain from remarking the moment when the last bottle of liquid hydrogen was let into the apparatus: if helium was not liquefied now, it would be some time before the stores were replaced and there could be another attempt.

Raising and then lowering the pressure to 75 atmospheres produced a "remarkably constant" reading of the thermometer at 5 K. It was 7:00 in the evening, and yet nothing could be seen in the glass receptacle. Schreinmakers suggested to Onnes that the refusal of the thermometer to budge from the 5 K reading was similar to what would occur "if the thermometer was placed in a liquid." Going beneath the vessel with an electric light, Onnes peered up at it, and saw clearly the outline of a liquid in the vessel, "pierced by the two wires of the thermoelement." "It was a wonderful moment," he later remembered: the surface of the liquid helium "stood out sharply defined like the edge of a knife against a glass wall." He had liquefied helium. His only regret, just then, was that he could not show liquefied helium to his friend van der Waals, "whose theory has been guide in the liquefaction up to the end." That gratification would have to wait a few days, since van der Waals was in Amsterdam, not in the small crowd of observers at the laboratory.

Though the liquefaction of helium had been predicted, it was nonetheless a spectacular achievement, a triumph of science harnessed to technology. In one last leap that built on all of the preceding ones, Onnes had lowered temperatures to a point scientists believed approximated the conditions of interstellar space, a point very near to a physical limitation of matter. Absolute zero lay ahead, but there was growing doubt that it could ever be reached, though not because of a lack of technological prowess.

Onnes sent a telegram announcing the liquefaction of helium to Dewar; it bore the wrong date, July 9, a date Dewar would repeat in a footnote to an article he was then in the process of composing.

Dewar's response encapsulated his complicated feelings at this event:

CONGRATULATIONS GLAD MY ANTICIPATIONS OF THE POSSIBILITY OF THE ACHIEVEMENT BY KNOWN METHODS CONFIRMED MY HELIUM WORK ARRESTED BY ILL HEALTH BUT HOPE TO CONTINUE LATER ON.

The Leiden team ran the machinery for two hours more, and then shut it down. "Not only had the apparatus been strained to the uttermost during this experiment and its preparation, but the utmost had also been demanded from my assistants." Onnes made certain in his *Communication No. 108* to express his "great indebtedness" to Flim for his "intelligent help" in having constructed the apparatus.

In that communication, written a few days after the event, Onnes recounted additional experiments he had conducted on the liquid helium, rapid attempts to discern its properties while he and his associates still had the liquid in the vacuum flask. He noted that several things about the liquid helium were what Dewar had predicted — the small surface tension, the difficulty of seeing the liquid, even the critical temperature, which Dewar had once postulated at 8 K and then revised to 5 K, and which was reached at 4.5 K. Dewar appeared to have been wrong only about the density of the liquid in relation to the saturated vapor; it was eleven times as dense, not seventeen times.

Beyond those expected parameters, there were some strange and inexplicable findings. Liquid helium had unprecedented low surface tension. Onnes believed that liquid helium's most astonishing property was its density, eight times lighter than that of water, which he thought might account for liquid helium's low surface tension — but then, hydrogen was also lighter than water, and its surface tension could be felt. A second curious finding was the failure of the liquid helium to solidify when cooled further by the same techniques Dewar had used to solidify hydrogen. No current theory could account for either of these results. Nor had Onnes explored such areas as magnetism, electrical conductivity, and other properties of matter already known to be grandly affected in the nether regions of temperature.

For all these reasons, it became clear to Kamerlingh Onnes that, having reached the penultimate landmark of Frigor, the task of learning more about the behavior of matter in the ultracold environment was just beginning.

A Sudden and Profound Disappearance

WHEN JAMES DEWAR LEARNED that Heike Kamerlingh Onnes had liquefied helium, he upbraided his assistant Robert Lennox for failure to provide him with a good enough separator of helium gas from the Bath Springs sands so he could have reached that goal in advance of his rival. The two Scots quarreled, and Lennox walked out of the Royal Institution, vowing never to return until Dewar was dead.* The defection of Lennox was a heavy blow to Dewar's low-temperature research, and after Lennox's departure, Dewar turned back from his march toward the cold pole and he never made another attempt on absolute zero, dropping his work on liquefaction to pursue other scientific endeavors. He did, however, dutifully report Onnes's achievement in liquefying helium to a meeting of the BAAS; and when Elisabeth Onnes sent him a telegram saying that her husband had taken ill, he wrote back that he was "grieved to learn" that the "great" man was ill, "but not surprised after the strain of his epoch-making work." In time, Dewar consoled himself with the belief, as he put it in a letter to Onnes, that "we must not forget what we have done someone else

---

* True to his word, Lennox returned to the Royal Institution after Dewar's death in 1923.

might have done," provided they had the aptitude, the funding, and the resources to do the work.

Dewar was even sicker now than in past years; before long, he would be operated on for cancer of the vocal cords and would require some time to recover. Unwilling to leave low-temperature research behind completely, he let Onnes enlist him, along with Olszewski and Linde, in a new institute to draw up standards for the refrigeration industry and for laboratories working in the ultracold environment. Shortly, Onnes became known in the press as "the gentleman of absolute zero," but it must be recognized that Dewar, at least in his relations with the man who had beaten him to the liquefaction of helium, also behaved as a gentleman.

After the liquefaction of helium, Kamerlingh Onnes was virtually alone in the field. Only he possessed the ability to produce liquid helium, and although his process did not make much in each run, if he managed carefully there would be enough liquid helium for conducting experiments. The knowledge that he had something of a monopoly made him redirect the thrust of all four fields of physics at Leiden; from 1908 on, 75 percent of the research done in thermodynamics, electricity, magnetism, and optics used low temperatures as a tool. For his own research, the first experiments were to continue the drive toward the cold pole. Very soon, by pumping off helium vapor, he depressed the temperature of the liquid helium to within 1.04 K of absolute zero. However, when it seemed apparent that manipulating the pressure did not push the liquid helium into the solid state or further reduce the temperature, Onnes called a halt in the attempts to reach absolute zero and decided instead to use liquefied helium to test the properties of matter in the neighborhood of a few degrees above absolute zero.

Part of the reason for Onnes's abandoning his prior single-minded quest for the cold pole was that absolute zero was now agreed to be impossible to achieve. In 1905 Walther Nernst of Berlin had shown definitively that it was at an infinite distance.

The vague idea that absolute zero existed but was unreachable had been around for a while. Nernst solidified it by logically relating it to Rudolf Clausius's concept of entropy. Examining the results of liquid-hydrogen experiments, Nernst contended that what lowered the temperature was the extraction of heat by evaporation, which reduced the entropy of the liquid. Nernst argued that the total energy of a system was constant so long as it was isolated. But when the system interacted, the change in energy equaled the sum of the work done on it and the heat absorbed by it. This meant that entropy would be vanishing as the system's temperature approached absolute zero. This notion Nernst formulated into what came to be called the third law of thermodynamics: as temperature approaches absolute zero, entropy approaches a constant value, taken to equal zero. This was the opposite of Amontons's belief that at absolute zero, all *energy* would vanish; of course, Amontons wrote 150 years before Clausius invented the concept of entropy. Nernst's third law additionally held that absolute zero could never be reached because the closer it seemed, the more difficult it became to convert heat energy into entropy. In other words, absolute zero was at an infinite distance, and therefore unattainable.

The feeling that the cold pole was beyond reach refocused post-helium-liquefaction research, aiming it again at where Dewar and Fleming had left off in the late 1890s, the altered properties of matter at ultra-low temperatures. Among the most interesting one they had investigated was the deep drop in electrical resistance. A substance's electrical resistance is the degree to which it retards the passage of an electric current through it. Low resistance or "resistivity" means a substance is a good conductor; high resistance, a good insulator. As Onnes wrote, he and Jacob Clay undertook "to corroborate and extend earlier measurements by Dewar" on the decline of resistance at low temperatures.

On the basis of finding a steady decline of resistance at liquid-nitrogen temperature, and obtaining apparent confirmation of the rate of descent in their liquid-oxygen tests, Dewar and Fleming had predicted that at absolute zero, pure metals would be perfect conductors of electricity. Dewar had revised that prediction when readings taken at the still-lower liquid-hydrogen temperature did not fit. Certainly, resistance at the liquid-hydrogen level was appreciably lower than that at the liquid-oxygen level — Dewar beautifully demonstrated this in a 1900 lecture, first immersing a lamp with a copper resistance coil into liquid oxygen to show that the bath brightened its light, then removing the coil from the oxygen and reimmersing it in a lower-temperature liquid-hydrogen bath, with the result that the lamp's light shone more brilliantly. Dewar's calculations of the "downward slope of resistance" had initially suggested that the "resistivity" of a copper wire would drop to zero at −233°C (40 K), but it did not, and when at the liquid-hydrogen temperature of −253°C (20 K) it still had not dropped all the way to zero, he wrote that "we must infer that the curve co-relating resistance and temperature tends to become asymptotic at the lowest temperatures."

Dewar did not try to guess what might have caused the "asymptotic" resistivity readings. A year later, he changed his opinion on the disappearance of resistivity at absolute zero. Now he believed that below a certain temperature point, resistance levels might persist no matter how much further down the scale experimenters went.

Dropping resistance levels as temperatures seriously declined had been predicted in 1864 and documented in the 1880s by von Wróblewski and Olszewski — separately, of course. And so in the phase of exploration that Onnes pursued in the years immediately following his 1908 liquefaction of helium, it was natural for him to consider electrical resistance at liquid-helium temperatures as a main point of his research. But he also wanted to look into other properties of matter associated with liquid helium. Its low temperature could produce changes in the magnetization and in the specific-heat capacities of metals.

In trying to use liquid helium as a tool, Onnes ran into the same barrier that had once beset Dewar: the practical difficulty of handling the intensely cold liquid. The cryostats Dewar had invented to work with liquid oxygen proved inadequate for preserving liquid helium. Dewar had walked around a lecture platform carrying liquid hydrogen in an open vessel. That could not be done with liquid helium, because allowing even the slightest additional heat into a vessel could cause the liquid helium to again become a gas. Completely enclosed containers had to be made that could maintain the helium as a liquid while experiments were done with it. The process of creating the necessary equipment to handle liquid helium during experimentation took until 1911.* Only then could Onnes use liquid helium to critically test the properties of matter.

Electrical resistance at ultra-low temperatures took center stage, not least because Onnes saw in it a chance to resolve questions for which competing theories and theorists proposed very different answers. Several leading authorities had made guesses as to what would happen to the resistance of a good conductor as the temperature was brought down near absolute zero. Lord Kelvin was the most prominent proponent of a belief, held by many other scientists, that the "death of matter" would occur at absolute zero, and that this would mean infinitely large rather than infinitely small resistance. Kelvin argued that because the electrical resistance of copper was still measurable at 20 K, the temperature of liquid hydrogen, when it had been predicted to disappear at 40 K, it was probable that as the temperature of the copper was lowered even closer to absolute zero, the electrons of the copper would freeze into place; that freezing would produce an upward "spike" in resistance, an indication that at absolute zero, resistance would be infinitely large. In Kelvin's view, while resistance might pass through a mini-

---

* Liquid helium had to be transferred from the vessel in which it had been collected at the end of the liquefaction process, by means of a specially constructed siphon — cooled and isolated from the environment — and held in another vessel, one large enough to also contain measuring apparatus for whatever experiment was being conducted and a stirrer to ensure uniformity of temperature in the liquid helium.

mum on the way down the temperature scale, perhaps in the neighborhood of 10 K, when the temperature was further reduced, "electron condensation" would make resistance rise again.

For some time, Onnes agreed with Kelvin, possibly because the
theory that Kelvin formulated regarded the electrons in the conductive metal as a substance that could be described by a van der Waals
equation of state. But Onnes had to contend with the equally per-
suasive work of Nernst. The German physicist's theory implied that
the resistance of a pure metal would disappear completely at absolute zero. Nernst had visited Onnes at Leiden, and the two men
corresponded.

Onnes resolved his inability to choose between Nernst and Kelvin
by doing a few experiments using liquid helium; these showed that
the resistance of certain metals continued to drop or (in the case of
pure platinum) to remain constant approaching absolute zero. Concluding that the evidence "excite[d] a doubt of Lord Kelvin's opinion," Onnes then partially abandoned his belief in infinite resistance
at absolute zero. But he could not agree entirely with Nernst, either,
and so made up his own theory, based on the 1864 contention that at
the lowest temperatures, impurities in a metal would prevent resistance from disappearing entirely.*

Over the course of his many years of research on various subjects, and hundreds of experiments, Onnes fashioned dozens of
working hypotheses. The evidence he turned up disproved nearly
all of them. This does not indicate that Onnes was a poor theorist;
rather, it suggests an important ability to make educated guesses
and a willingness to toss them aside when they no longer seemed
viable, to search for better explanations of the experimental results. The capacity to let go of working hypotheses when they
proved inadequate was a reflection of Onnes's scientific worth and
integrity.

---

* The theory also had portions that made reference to the vibration of molecular particles;
more about this aspect later in the chapter.

To prove or disprove the hypothesis that the impurities in a metal kept its resistance from vanishing as the temperature reached a few degrees above absolute zero, Onnes gave up on platinum and began to experiment with "the only metal which one could hope to get into wires of a higher state of purity, viz. mercury." A fortuitous choice. The other metal he had been working with, in a very pure state, was gold, and had Onnes concentrated on gold and not mercury, he would not have obtained the same startling result. He could refine mercury at room temperature, and he did so repeatedly, until certain that he had removed all possible impurities.

In December 1910, Onnes was thrilled by the award of the Nobel Prize to his friend van der Waals, for his theoretical contributions. There were rumors that Onnes might be next in line, but there was also a feeling among some scientists that liquefying helium was simply a technological feat, not a discovery or theoretical advance considered worthy of a Nobel. Onnes shrugged off both notions, and continued working.

In April 1911, time became of the essence. Onnes learned from his journal reading that Nernst was obtaining preliminary results on the conductivity of metals at high and at low temperatures. Onnes was also concerned about Einstein, who was known to be investigating the elastic constants of metals at both high and low temperatures. Either of these formidable scientists could beat Onnes to the next accomplishment in low-temperature research.

It took until the summer for Onnes and his colleagues at Leiden — Flim, Gilles Holst, and Cornelius Dorsman — to refine their mercury to their satisfaction and to set up an experimental apparatus to test its electrical conductivity at very low temperatures. The mercury was held in a U-shaped tube with wires running out of both ends, from which they would measure the metal's electrical resistance; as the temperature was lowered by means of liquid helium, the mercury congealed to a solid. Onnes and Flim worked with the cryogenic apparatus in one room, while in a dark room more than 150 feet away, Holst and Dorsman sat and monitored the resistance readings taken by a galvanometer, an instrument that records minute changes in electrical current by means of a coil of wire surrounding a magnet. As the temperature was pushed below 20 K above absolute zero, the resistance continued to decline but slowed its pace of descent, with each 1-degree drop in temperature no longer matched by equivalent percentage drops in the resistance.

During these electrical-resistance experiments there was not the same aura of excitement and expectation as had suffused the Kamerlingh Onnes laboratory three years earlier. Back then, Onnes had had the equivalent of a detailed treasure map in hand — composed of van der Waals's theory and Onnes's own isotherms — to guide him to the liquefaction of helium. Part of the thrill of that discovery was finding that the treasure was located precisely where the map had predicted it would be. In mid-1911, while Onnes and his colleagues did have a map of sorts — his hypothesis that the resistance of mercury would be prevented from vanishing at a few degrees above absolute zero because of impurities in the metal — they recognized that other maps pointed to different locations for the treasure. Suspecting that all the maps could well be wrong, they no longer labored in expectation of knowing just when a big event would take place or what it would consist of. So the thrill, when it came, was the unforeseen nature of what they discovered in the furthest region of the country of the cold.

As the mercury reached the temperature of 4.19 K, the electrical resistance of the mercury solid fell abruptly — as though it had been driven off a cliff — to a level so low that the galvanometer no longer registered any resistance to the current. At 4.19 K, mercury's electrical resistance just disappeared. Not confident in the accuracy of this result, Onnes tried the experiment over and over again. Every time, the findings were the same: no resistance at 4.19 K above absolute zero. Onnes and his colleagues then assumed that their apparatus might be subject to a short circuit and took a few days to replace the U-shaped tube with a W-shaped one that had electrodes extruding from all five points, which gave them more places between which to measure resistivity. Even with this more sensitive setup, when the temperature of the apparatus was lowered to 4.19 K, the resistance readings on the galvanometer fell to zero.

Flim later told a physicist who joined the laboratory in the 1930s that during the 1911 experiments, one of the blue boys was assigned to maintain pressure in the apparatus; however, because watching a dial was a very boring job, in the course of one run the young man fell asleep, and when the dial began to move, he did not see it and so did not alert his superiors to properly adjust the apparatus. The pressure dropped, the temperature in the apparatus rose above 4.2 K — and in the galvanometer room, Onnes's colleague Holst saw the resistance reading suddenly jump into the measurable range.

This last, reverse demonstration of the transition that seemed to be occurring at 4.19 K may have been the one that finally convinced Onnes that he had discovered a novel property of matter at extreme temperatures, a property he did not even name in his first articles, including that entitled *On the Sudden Rate at Which the Resistance of Mercury Disappears.* But he did call attention to the astonishing fact that in this state, "the specific resistance of a circuit becomes a million times smaller than the best conductors at ordinary [room] temperatures."

Onnes initially thought the drop in mercury's resistivity at 4.19 K confirmed his theory about resistance being linked to the purity of the metal, but the steepness of the cliff over which the resistance fell showed him he had been wrong. Now he was at a loss as to how to explain the "disappearance" of the resistance, and equally perplexed by the results of experiments on tin and lead, where the resistance also dropped abruptly, and on gold, where it did not. Physicist Kurt Mendelssohn later suggested that Onnes's puzzlement was reflected in his dearth of published articles about the phenomenon in 1912, the year after he had first announced and correctly described the drop-in-resistance phenomenon, a time in which Onnes continued to conduct experiments. An equally likely reason may have been Onnes's deteriorating health, which increasingly kept him confined to his home and bed. It was not until his second paper of 1913 that he used the word *supraconductivity* to describe the phenomenon, a term he later discarded in favor of *superconductivity*.

Onnes may have been cautious because his peers did not initially recognize superconductivity as of great importance. At a conference in 1912, when Onnes reported on the discovery, his audience did not show much interest; only two questions arose on the subject. Moreover, when James Dewar — who had pioneered work on electrical resistance at low temperatures — heard about superconductivity, he made no comment that has been recorded, and he did not send Onnes a telegram of congratulations, as he had done several times earlier. Perhaps Dewar, too, did not realize the magnitude of the discovery.

As for Kamerlingh Onnes, though he might not have understood right away all he had accomplished, he did envision a future for superconductivity, and by 1912 he had constructed an experiment to provide fodder for it: he introduced a current into a superconducting circuit he had formed into the shape of a ring, then removed the battery that had generated the current. Inside the ring, the current continued to run, and run, and run, with no measurable change in intensity. Years later, one leading physicist who visited the lab wrote a letter about this demonstration to another, saying, "It is uncanny to see the influence of these 'permanent' currents on a magnetic needle. You can feel almost tangibly how the ring of electrons in the wire turns around, around, around — slowly and almost without friction." Max Planck, inventor of the quantum theory that was about to revolutionize physics, was also extremely interested in what Onnes had discovered. And Onnes himself would shortly predict that someday superconducting wires would enable human society to transport electricity in ways much more efficient than those then in use: electrical power plants could be situated hundreds of miles away from the places where most power was to be used; transmission costs would drop precipitously, lowering the cost of electricity to its users; and the world would have a virtually unlimited supply of electric power.

This was a good dream. But reality soon intruded. As Onnes continued to experiment in the ultracold environment, he tried to determine what effect magnetism would have on materials at very low temperatures, and found — to his dismay — that a magnetic field of a few hundred gauss, the strength exhibited by an ordinary household magnet, was enough to eliminate the superconductive state in materials such as mercury, tin, and lead. The moment a magnetic field was turned on in the vicinity of the material that had been rendered superconductive by liquid helium, the superconductive state appeared to vanish. This seemed to mean it would never be possible to have superconducting wires that would revolutionize the use of electrical power in the world.

This almost immediate dashing of his dream, and the inability of other contemporary scientists to realize the importance of superconductivity just then, may help to explain why, when the Nobel committee awarded Heike Kamerlingh Onnes the 1913 prize in physics, it cited the seventy-year-old scientist "for his investigations on the properties of substances at low temperatures, which investigations, among other things, have led to the liquefaction of helium," and did not specifically mention his discovery of superconductivity.

In his speech accepting the Nobel Prize in Stockholm, Onnes offered no philosophic ruminations on how the world had changed because of his discoveries. Rather, he treated the Nobel address as though it were a routine though nostalgic lecture to a scientific conference. He reported on the "Leiden und Freuden," the disappointments and joys, not only of his own research but also that of Dewar, Olszewski, von Wróblewski, Pictet, Cailletet, and Linde over the past thirty-five years. He recounted in detail the events of July 10, 1908, when helium had first been liquefied, and his wish to have shown the liquid helium immediately to van der Waals. Onnes made certain to mention superconductivity, expressing again his wonder that "the disappearance [of electrical resistance] did not take place gradually but *abruptly*," an occurrence that "brought quite a revelation." In a similar awestruck manner, Onnes detailed his findings about the extremely low density of helium. Explanations for these phenomena had still not been made, and in a fervent prediction, Onnes suggested to the Nobel audience that when explanations for these strange phenomena were made, they "could possibly be connected with the quantum theory."

In a way, this was Onnes's acknowledgment of a most important point that could not have been understood even a few years earlier: that the liquefaction of helium and the discovery of superconductivity were the last triumphs of what would shortly be referred to as "classical physics," the physics of Newton and Boyle, Kelvin and Clausius, the physics of the past. Classical physics described objects and their motion, while quantum physics described matter *only* in terms of motion, wave motion. More so than many in his age group, Onnes understood and accepted that a new generation of scientists — almost a new breed — able to embrace supremely sophisticated, complex, counterintuitive ideas, were in the process of supplanting the generation that he and James Dewar so well exemplified. In other articles composed around this time, he reiterated his belief in the quantum theory of Planck, asserting that it might provide the "mechanism" responsible for the disappearance of electrical resistance in the several superconductors discovered by Onnes to exist at a few degrees above absolute zero.

To understand why Onnes would predict the ability of quantum theory to explain superconductivity, we must backtrack to 1907, a year before he liquefied helium. In that year, an examination of the "specific heat" of copper at extremely cold temperatures led to the solution of one set of troubling anomalies previously highlighted by near-absolute-zero research, and in the process, produced an important verification of the quantum theory that Planck had articulated in 1900. The problem solver was Albert Einstein.

Specific heat had fascinated physicists and chemists since 1819. That year, French chemists Pierre-Louis Dulong and Alexis-Thérèse Petit defined it as a measure of the heat required to raise the temperature of a small quantity of a substance by a fraction of 1 degree and determined the specific heat of all sorts of materials. They produced a law, an equation that accurately predicted the specific-heat capacity of common materials such as lead and copper. But by 1875 — that is, before Cailletet and Pictet liquefied nitrogen and oxygen — it had already become obvious from research in the region just below 0°C that the Dulong-Petit law did not hold for all temperatures. The situation was similar to what Andrews and van der Waals had encountered when dealing with Boyle's law: an equation that explained things quite well at room temperature proved untenable when the temperature was dropped well below the freezing mark. When liquid hydrogen became available, Dewar, Onnes, Olszewski, and other researchers noted that in the temperature region it made accessible, about 20 K, the specific-heat capacity of copper dropped to a mere 3 percent of what it was at room temperature. The problem that still remained in regard to Dulong-Petit was to come up with an explanation, and an equation, incorporating that old law's description of how matter behaved at normal temperatures but also encompassing how matter behaved at the newly reached lower temperatures.

This was just the sort of problem Einstein liked to tackle. Dulong and Petit had based their law on the "equipartition of energy," describing the vibration of individual atoms. Einstein realized he had to replace their description with one that took into account the "quantization" of the atoms' vibrations. In the picture of thermal motion that had evolved by 1907, an atom was considered to be an oscillator with six "degrees of freedom," each one containing some energy. By then, Einstein had decided that Max Planck's work on "quanta," the small parcels into which many forms of energy are subdivided — work that Einstein had originally thought was in conflict with his own — was really complementary to his own. So he extended Planck's quantum theory, arguing that atomic vibrations were quantized, meaning that the atoms did not vibrate freely but in small, measurable, incremental steps. That was the solution to the specific-heat puzzle. As the temperature of the copper fell, Einstein suggested, more and more of its atoms were constrained from vibrating, leading to an exponential drop in the metal's specific-heat capacity. He wrote an equation that matched reasonably well — not perfectly, but fairly closely — the observed data for the specific heat of copper, all the way from the Dulong-Petit area of 80°F, down through the liquid-oxygen and liquid-hydrogen temperatures, to around 10 K. Not only did his equation predict the changes in specific heat as temperature fell, but by showing that quantum theory could explain something that had previously been beyond understanding, Einstein's proof also upheld the insight of Nernst's third law of thermodynamics and provided an early verification of the truth and worth of Planck's quantum theory.

Einstein's successful explanation of the drop in specific heats had excited Onnes well before his discovery of superconductivity. At the outset of his resistance experiments, Onnes had cobbled together a working hypothesis that was a hodgepodge of classical and quantum ideas, combining, in addition to the business about the impurities in the metal, the equations of state of van der Waals, married to Planck's notions of vibrating particles. Then Onnes's experiments with mercury revealed the sudden fall in resistance at 4.19 K — a result, he wrote with characteristic aplomb, "not foreseen by the vibrator theory of resistance that I had framed." So he had to abandon that hypothesis as he had abandoned others, but he maintained his belief that eventually quantum physics would provide the key to understanding superconductivity. And when the English physicist J. J. Thomson — later a Nobel Prize winner himself — postulated an explanation for superconductivity that did not include quantum theory, Onnes went out of his way to reject it publicly, on just that basis.

While Onnes had been on the glide toward his Nobel, between 1908 and 1913, Dewar had not faded graciously away. In 1911 he commissioned the refitting of the amphitheater at the Royal Institution, at his own expense, in celebration of his having held the Fullerian Chair of Chemistry even longer than Faraday. In his research after 1908, Dewar made several important contributions, among them the invention of a charcoal-based calorimeter, which he used to measure the heat capacities of many elements and compounds in the liquid-hydrogen range and below. In 1913 he discovered that at 50 K, the heat capacities of the solid elements were related to their atomic weights by a logarithmic equation. He also returned to several other matters that had intrigued him in earlier years, among them soap bubbles and thin films, on which he now did some important research, and explosives, building on his pioneering work with charcoal. Back in 1889 Dewar and Sir Frederick Abel had invented cordite, a gelatinized mixture of nitrocellulose and nitroglycerin used as a smokeless explosive.

Regarding explosives, Dewar came to believe that some of the innovations he had introduced had been purloined, without credit or payment, by Alfred Nobel and his heirs, and he brought suit against them. The suit was eventually dismissed as having no merit.

Dewar never received a Nobel Prize for his research, although his liquefaction of hydrogen had been the key experiment in the descent toward absolute zero, and although his invention of the cryostat was essential in all experiments conducted at ultra-low temperatures. He had no pure discoveries to his name, and no theories, and Nobels usually went to discoverers and theorists. There may also have been resentment against Dewar among the heirs of Alfred Nobel responsible for the administration of the prizes, though the recipients were always chosen by a committee of experts in the field; as for that, Dewar's confrontational style with Rayleigh, Ramsay, and Travers had also earned him black marks among the better-respected English chemists and physicists of the day. In the elaborate procedure of nomination for the Nobel, their overt support would have been necessary to put him on the final ballot.

In August 1914 the Great War began, pitting the forces of countries from the British Isles to the Balkans against one another. Among the early collateral-damage casualties was Olszewski. Austrian soldiers invaded the building in which his laboratory and quarters were located and turned it into their dormitory. Already frail, Olszewski took to his bed; ever the scientist, on the night that death neared, he noted down its approaching symptoms, sandwiching the observations between his requests for funeral arrangements.

Another casualty of the war was low-temperature research. The still-small supply of helium was conscripted for the military in the combatant countries, which started to use helium gas for dirigibles and lighter-than-air espionage and antiaircraft balloons.

And so the exploration of the country of the cold came to a temporary halt at the discovery of superconductivity, that first indication of the profound transformations of matter that the ultracold environment could produce. This was not using cold to make eggs glow in the dark — it was far more basic and interesting. Everyone hoped ultracold research would resume after the war, because so many things were yet to be learned, first among them the explanation of why the superconducting state was brought into existence at low temperatures. Until that resumption, the discovery of superconductivity was a beacon lit at a very far outpost of Frigor, a fitting fulfillment of the scientific explorers' long quest into its frigid realm. In their race toward absolute zero, the generation of Onnes, Dewar, Olszewski, von Wróblewski, Cailletet, Pictet, Linde, and

Hampson had successfully explored a difficult field, and they had bequeathed to the next generation exciting and formidable tasks reminiscent of those that had faced the theorists of Salomon's House in Bacon's fable: to distill the "knowledge of Causes, and the secret motion of things," and to use what they learned for "the enlarging of the bounds of Human Empire, to the effecting of all things possible."

12

Three Puzzles
and a Solution

HAT KAMERLINGH ONNES AND his fellow turn-of-the-century researchers did not immediately realize, in the period before the onset of the Great War, was that in their work on the ultracold they had unlocked a treasure chest of information about previously unknown aspects dealing with the operation of the normal world as well as with that strange one in the vicinity of absolute zero. The baubles in this trove would eventually provide avenues of understanding to the primal secrets of the universe.

The single most significant roadblock to reaping those understandings was the enigma of superconductivity, whose solution would take another sixty years and require the efforts of battalions of good scientists. During the height of the attempts to figure it out, Felix Bloch coined what he called an axiom: "Every theory of superconductivity can be proved wrong." For many years, that was the only correct statement in the field.

A more appropriate saw would have been that in science, each new discovery raises more questions than it answers. Chemist Leo Dana, fresh from receiving his doctorate at Harvard in 1922, ran right into one of those new questions when he arrived at Leiden to spend a postdoctoral year with Onnes.

Unfortunately for Dana, the day of his arrival was the day after the death of Onnes's intended successor, J. P. Kuenen. The laboratory was in shock. Among other reasons for the dismay, Dana learned, a battle likely to further upset the laboratory would now take place between Protestants and non-Protestants for the position of heir apparent; for centuries, Leiden had been a Protestant university. A compromise shortly saddled Wilhelmus H. Keesom, a Catholic, with a Protestant codirector for a time.

Dana busied himself with learning the ropes. His questions had to do with Onnes's postwar research into the unusual density of liquid helium. At the boiling point, 4.2 K, it rose dramatically from what it had been at warmer temperatures and passed through a maximum at 2.2 K, but it gradually declined thereafter, even when the temperature was dropped to less than 1 degree above absolute zero. What accounted for this peak and change in density? Could the density of helium be tied to the onset of superconductivity in metals alone or in conjunction with the effect magnetization had on superconductivity?

Dana wanted to work with Onnes on these questions. The director was now an old and sick man. He seldom visited the laboratory, though he maintained a correspondence with nearly everyone of importance in physics, from Einstein to W. C. Röntgen to lesser-known researchers in the low-temperature laboratories in the Soviet Union, and he worked assiduously to help build up the commercial liquefaction and refrigeration industries of the world. Onnes communicated with colleagues principally by telephone; emphysema made it difficult for him to speak, but when he did, a French correspondent later recalled, his brief comments were always on target and often insightful.

Dana was invited to Ter Wetering. Its appearance, he later recalled, made it "evident" that Onnes was now a wealthy man. "I was ushered into his study, furnished with antique furniture, oriental rugs and paintings; looking out the window, one saw the lovely scenery of the Dutch countryside. He was dressed in a fancy velvet gown — the typical man of means." Onnes advised the young Harvard-trained chemist, "If you ever see a ripe plum on a tree, reach up and grab it."

In the laboratory, Onnes put Dana to work in trying to grab hold of something they both thought of as a ripe plum, the latent heat of liquid helium as it vaporized. What Onnes and Dana together found in their investigations of this phenomenon they labeled "remarkable." "Near the maximum density," they wrote in an article, "something happens to the helium, which within a small temperature range is even discontinuous."

It was discontinuous in the same remarkable way as the "abrupt" drop in resistivity when mercury became superconducting. Next they found a similar change in the specific heat of liquid helium, at the same transition temperature they'd identified in the latent-heat experiment. When the temperature dropped below that line, the specific heat became much larger than expected, or than predicted by any theory, including Einstein's. Onnes did not want to report the figures, because they were too large and because they contradicted Einstein's work. The joint paper Dana yearned to have published had to wait for several years.

Onnes and Dana could not figure out why the specific heat of the helium changed so markedly, but they determined the precise point at which the discontinuity started, 2.2 K, which a visitor to Leiden, Paul Ehrenfest, labeled the "lambda point," because the shape of the curve describing the specific heat resembled the form of the Greek letter lambda. Identifying the lambda point, however, did not mean anyone could yet understand the "something" happening to helium at 2.2 K.

His year of postgraduate study almost up, Dana made ready to return to the United States and was invited to a farewell dinner at Ter Wetering. One moment stuck with him: at the dinner table, when Elisabeth Onnes wanted to have the next course served, rather than summon the waiters herself, she followed tradition and asked her husband to ring the bell; with great difficulty, Onnes got out of his chair and walked to a side panel to pull the bell to summon them.

For Onnes, 1923 was a year of great losses — the deaths of van der Waals and Dewar. "One of the great figures of modern physics and physical chemistry," Onnes described his friend van der Waals, who died at age eighty-five, in an obituary in *Nature,* expressing admiration for his "severe culture of the ideal" and repeating Dewar's characterization of van der Waals as "the master of us all." The theorist had begun to deteriorate in 1913, Onnes wrote, and "[a]t last, only short visits allowed us to show to the venerated and beloved friend, whose heart we felt unchanged, what he had done for us."

Nineteen days after van der Waals's death, Sir James Dewar died, on March 27, 1923. A decade had elapsed since Dewar had first begun to reveal to Onnes in letters his bemused amazement that, though ill most of his life, he had actually reached the age of seventy. By 1923 Dewar was eighty-one, and he had continued to do important work on films and with a charcoal-gas thermoscope he constructed to measure infrared radiation. From the basement of the Royal Institution, where he had conducted his low-temperature experiments, Dewar had moved up to the attic, whence he measured the radiation from the sky. He took his last readings just a few nights before he was confined to his bed by his final illness.

Onnes survived another three years, becoming less able to draw breath with each passing week. At his death in 1926, he was mourned throughout the world. The outpouring accompanying Onnes's passing was greater than for Dewar, a loner who left no school of successors, since the heirs of Onnes were everywhere in the laboratories of Europe and the British Isles.

Just three weeks after Onnes died, his last collaborator, W. H. Keesom, completed what Onnes had worked toward for fifteen years, the solidification of helium. Because helium seemed to remain liquid as far down toward absolute zero as Keesom could reach while keeping the pressure moderate, he was able to make crystals form in the helium only by applying greater amounts of external pressure.

The solidification of helium led to Keesom's refining of the Onnes-Dana data on specific heats. Keesom found that while liquid helium boiled at 4.2 K, when it descended to the lambda point of 2.2 K the boiling ceased, the bubbles stopped, and the liquid helium became completely still. These dramatic shifts at the lambda point suggested to Keesom that the liquid from 4.2 K down to 2.2 K must be treated as a distinct phase called helium "I," while the liquid below 2.2 K was very different and should be regarded as another separate phase called helium "II." Compared with helium I, helium II had a smaller density, a greater heat of vaporization, and a smaller surface tension.

Scientists were generating additional questions about the behavior of atoms in the vicinity of absolute zero, based on the possibilities raised by Walther Nernst's third law of thermodynamics. If Nernst was correct, as one approached ever closer to absolute zero, the atoms ought to increasingly align themselves in a formation approaching perfect order. In 1925 Einstein turned his thoughts once again to this area of inquiry, spurred by the work of Satyendra Nath Bose, an Indian physicist. As atoms slowed down and approached a virtual standstill, Einstein argued, they would be close enough together to cause their wave functions to overlap, merge, and cooperate, producing a state of matter unlike any already known. This hypothetical new phase or state of matter came to be labeled the Bose-Einstein condensate, and during the next seventy years, physicists would try unsuccessfully to create it to prove Einstein's contention.

Nernst's perfect-order idea, as refined by Planck and others to suggest that entropy measured the randomness of the microscopic state of a solid or liquid, also informed the separate inquiries of two other physicists, Dutch-born German theorist Pieter Debye and Canadian-born physical chemist William F. Giauque, newly appointed to the University of California at Berkeley faculty. It was understood, thanks to the work of Pierre Curie, that the magnetic susceptibility of a substance is inversely proportional to the absolute temperature — that at low temperatures, materials are more readily magnetized. It was also understood that when a material was magnetized, the magnetizing field worked on those of its atoms known as magnetic ions, turning them to face all the same way. Making them do so produced heat energy.

Near the end of 1925, Debye asked rhetorically "whether an effort should be made to use such a process in approaching absolute zero" and concluded that someone ought to do experiments to prove or disprove the theory. Giauque proposed the same process at virtually the same moment in time, but he wasn't content to stop at theory; he tried to construct an apparatus to achieve the goal.

Giauque magnetized a weakly magnetic salt at liquid helium temperatures, then surrounded it with a vacuum and demagnetized it, which changed the material's molecules from being highly ordered to being highly disordered. Doing that removed heat energy from the salt, which caused its temperature to fall. Giauque took another fifteen years to fully solve the technical problems associated with this process, which included finding a thermometer that could register within a few thousandths of a degree of absolute zero, but in 1938 he was rewarded by achieving the lowest temperature ever recorded up until that time, 0.004 K.

The demagnetization process represented an entirely new way of effecting low temperatures — beyond liquefaction, beyond Joule-Thomson expansion, beyond pressure. With it, scientists were essentially manipulating subatomic structures to produce cold. Giauque would receive the Nobel Prize in 1949 for chemistry; "adiabatic demagnetization" was among his contributions cited. Many practical applications have resulted from the ability to produce very low temperatures.

Meanwhile, work continued on superconductivity. Was the phenomenon specific to a few metals or common to many? Onnes and

his successors had demonstrated superconductivity only in relatively soft metals that had low melting points; in a Berlin laboratory,
Walter Meissner determined that some metals among the harder
group, such as the rare metals niobium and titanium, could be
induced to become superconducting. Later on, using the new technique of magnetic cooling, Meissner was able to show that other
metals — aluminum, cadmium, and zinc — became superconductors at temperatures below 1 K.

A geographic explorer coming across a form of life never seen
before, one that could be either a plant or an animal, has to decide
how to describe and investigate its properties. Treating it as a plant
mandates one line of inquiry; as an animal, another. As more and
more metals — but not all metals — were shown to be superconductors, an analogous basic question arose: Was the vanishing of electrical resistance due to the microscopic properties of the substance, a
change to the electrons or to the nucleus, or was the onset of superconductivity a change of thermodynamic state similar to the change
from a gas to a liquid? In the late 1920s, the betting favored a
microscopic-properties change, partly because no one other than
Einstein and Bose had been able to describe mathematically a state
of matter beyond those of gas, liquid, and solid.

When Kelvin and Clausius had written of states of matter while
constructing the laws of thermodynamics in the 1850s, they had
done so in terms of pressure, volume, and temperature. Van der
Waals in the 1870s had added molecular density as a descriptive. By
the late 1920s, yet another factor was thought relevant: a substance's
degree of magnetization. Taking into consideration the growing evidence that superconductivity and magnetism were related, Meissner
and his graduate student Robert Ochsenfeld decided to investigate
whether the change in a material as it became superconducting was
accompanied by a change in its ability to become magnetized — or,
as they put it in technical terms, by the degree to which magnetization permeated the substance.

Metals such as tin and lead could be readily magnetized at normal temperatures. Would that ability change as the metal was cooled down to the temperature at which it became a superconductor? Meissner and Ochsenfeld cooled two adjacent long cylinders of single crystals of tin and at the same time introduced a magnetic field. Just at the moment that the solid tin became a superconductor, they removed the external magnetic field, then took readings of the cylinders' residual magnetism. They found there was none. The metal had, in effect, expelled all traces of a magnetic field from its interior.

Shades of Faraday! Ninety years earlier, Faraday had investigated the magnetic properties of all sorts of materials — metals, carrots, apples, meat — and found that all of them possessed, to a small degree, a property he labeled diamagnetism. Now Meissner and Ochsenfeld had shown that a solid crystal of tin could be perfectly diamagnetic, expel magnetism totally, just as Onnes had shown that supercooling mercury wires to 4.19 K totally eliminated their electrical resistance. It was now clear that superconductivity was more than an anomaly in a substance's ability to conduct electricity — it was a phenomenon that also had to do with changes in a substance's magnetic susceptibility.

This second instance of the vast transformative powers of the country of extreme cold, called *superdiamagnetism,* was also a significant puzzle whose solution would take many years. But, its identification in 1933 seemed to insist that the onset of superconductivity might best be considered akin to a thermodynamic change of state similar to what happened when a gas became a liquid or a liquid changed to a solid. How and why this change occurred, no one yet knew.

Theories to explain the how and why kept cropping up at a rate estimated by Kurt Mendelssohn of several each year; most were soon dismissed because they did not explain both superconductivity and superdiamagnetism. Between 1933 and 1935, however, several sets of scientists made good guesses about the nature of superconductivity that included possible explanations of superdiamagnetism.

In 1934 C. J. Gorter and H. B. G. Casimir, colleagues and successors of Keesom at the Leiden laboratory, suggested a model for superconductivity in which two fluids of electrons existed simultaneously. One was an ordered, condensed fluid of the sort Bose and Einstein had thought about, with zero entropy, which meant it could not transport heat (the product of resistance); this they called the "superfluid." The other was composed of electrons that behaved normally. When the temperature of helium was lowered beneath the transition point, more of the electrons entered the superfluid state, and that change, Gorter and Casimir postulated, was what caused superconductivity.

Picking up on the two-fluids idea, the brothers Fritz and Heinz London, at Oxford — where they had fled after escaping the Nazis in their native Germany — theorized how a superconductor might produce superdiamagnetism. In his earlier doctoral thesis, Heinz had figured the depth to which a current on the surface of a superconductor penetrates to the interior of the metal. All currents coursing on the surface of a metal produce a magnetic field. Fritz used this fact to explain the Meissner effect (exclusion of a magnetic field from the interior of a superconductor), by showing that when the current on the surface of the superconductor partially penetrated the surface to produce a magnetic field, that surface field canceled out the already existing field, so that the interior of the superconductor remained field-free, that is, it had perfect diamagnetism. Based on his previous research charting the depth of penetration, Heinz could predict the shape of the curve describing the falloff of the magnetic field within a superconductor, and he could express it in terms of the number and density of the superconducting electrons.

At a seminal meeting of low-temperature researchers from many countries, held at the Royal Society in 1935, Fritz London summed up all the post-1911 theorizing by asking the scientists to take a dazzling imaginative leap — to stop thinking about superconductivity in terms of yesterday's classical physics and to instead consider superconductivity solely in terms of quantum physics and wave motions. Conceive of a superconductor, London pleaded, not as a collection of unrelated atoms but as one huge atom — and the problem will be more easily attacked. The big atom's interior order — the pattern synonymous with superconductivity — could be described, London said, by a single wave function. In other words, the superconducting state had to be produced by the electrons of this giant atom behaving coherently, or in unison.

Just when this notion was being put forth, a third ultra-low-temperature puzzle was discovered by a leading physicist unable to attend the meeting in 1935 at the Royal Society, though he desperately wanted to be there: Pyotr Kapitsa.

Previously, the Russian émigré had been the director of the Mond Laboratory in Cambridge, using equipment built expressly for him there at the suggestion of his teacher and mentor, Ernest Rutherford. The British had gone to this trouble because Kapitsa had demonstrated theoretical brilliance and practical ingenuity. For instance, he had invented his own apparatus for liquefying helium, which produced two liters an hour by an expansion process and did not require the cumbersome equipment of the five steps of liquefaction. By 1930 Kapitsa had accomplished so much that he had been elected a Fellow of the Royal Society, the first foreigner so honored in the previous two hundred years, and was doing important research on both magnetism and low temperatures. Each year he would take a trip back to Russia with his wife to visit their relatives, but when he did so in 1934, the couple were prevented from returning to Great Britain.

Two years of negotiating ensued until, in 1936, the Kapitsas, the Soviet government, and Rutherford completed a three-way deal. Kapitsa's wife agreed to go briefly to England to fetch the couple's two children and bring them back to the Soviet Union with her. In exchange for having his family reunited, Kapitsa accepted the directorship of a new laboratory in the U.S.S.R., and Rutherford arranged for the Mond's low-temperature-research equipment to be shipped to Kapitsa for use at the Institute for Physical Problems.

Once his machinery had arrived, in very short order Kapitsa succeeded in identifying and describing a third, new, and entirely unexpected aspect of matter in the region of ultracold. The discovery was galvanized into existence by a paper written by W. H. Keesom and his daughter Anna Petronella, which suggested that at the lambda point, the thermal conductivity of helium II increased over that of helium I by a factor of 3 million. This meant that helium II became a better conductor of heat than copper or silver, the best normal-temperature metallic conductors of heat.

Kapitsa, fascinated by helium II, used his imagination to make sense out of some odd things happening in research labs. Helium II had not been behaving like all other earthly liquids. It had escaped from containers dense and impermeable enough to prevent the leakage of any other fluid, even helium I. This ability of helium II had resulted in contamination of other fluids, making a shambles of experiments in several laboratories. Also, if a container of helium II was placed in a bath of helium II and filled to a level higher than the bath, the levels inside and outside the container would gradually equalize. Creeping up and over walls, defying friction and gravity, it seemed to refuse to adhere to normal physical rules of flow. "Helium," Kapitsa later wrote, "moves faster than a bullet."

Taking these Houdini-like effects as his starting point, Kapitsa tried to determine the parameters of helium II's escape artistry. Researchers at Cambridge and Leiden were also working on the problem, and the three groups kept in touch with one another by mail, telephone, and personal meetings, carrying on what Kurt Mendelssohn characterized as "ruthlessly searching discussions into the validity of each other's methods." In articles published beginning in 1938, Kapitsa seized the theoretical high ground and made order out of what had been perplexing chaos by describing what helium II was doing as exhibiting *superfluidity*.* He attributed superfluidity to changes in the viscosity that were intimately related to what had initially prevented Onnes and Dana from publishing their data in 1922: the sharp rise in helium II's specific heat, later identified by Keesom and Petronella as being 3 million times greater than that of helium I. This fantastic ability of helium II to conduct heat and its ability to move about as though nothing could stand in its way, Kapitsa suggested, were aspects of the same phenomenon.

He devised an experiment that demonstrated the effects of these behaviors. Inside a large dewar of liquid helium II, he placed a smaller one also filled with liquid helium, and in that "bulblet" Kapitsa inserted a capillary tube with one end sealed inside and the other open to the helium vapor. The outer dewar was necessary to keep the inner one at the proper low temperature. Kapitsa set up a weathervane sort of instrument near the open end of the capillary and applied heat to the bottom of the bulblet. A submerged jet of invisible liquid helium issued from the top and turned the weathervane. The experiment went on for hours, with the vane spinning and the bulblet of liquid helium as full at the end as it had been before the start. Kapitsa figured out that the heat transformed some of the superfluid to normal fluid, which produced the submerged jet. He concluded that helium II had no entropy and a viscosity 10,000 times lower than that of liquid hydrogen, that is, an almost unmeasurably small viscosity, virtually none at all.

"At first sight," wrote Russian physicist Lev Landau of Kapitsa's weathervane experiment, liquid helium's properties "seem completely absurd, like the story of the giraffe which evoked the exclamation, 'There ain't no such animal!'"

No viscosity.

No inner magnetic field.

---

* In the same issue of *Nature* as Kapitsa's first letter about the flow of helium II was a similar and equally revelatory letter on the same subject by John Allen and his Cambridge graduate student Donald Misener.

No electrical resistance.

The trio of unusual phenomena at the far edge of the ultracold was complete: superconductivity, superdiamagnetism, and superfluidity.

The discovery of this trio of phenomena meant the final eclipse of the old clockwork universe that obeyed Newtonian laws of motion. No adequate explanation of the new phenomena could be made by means of the old laws. Fortunately, though, this trio of puzzles surfaced at a time when quantum physics had matured enough to provide cogent attempts to explain superconductivity, superfluidity, and superdiamagnetism. Kapitsa was fond of saying that trying to detect the quantum nature of physical processes at room temperature was like trying to investigate the physical laws governing the collision of billiard balls on a table aboard a ship going through rough seas. And Landau would explain to his classes the advantages of lowering temperatures into the arena of the ultracold, where all sorts of processes slowed down and became more amenable to study. "As the temperature falls," Landau said, "the energy of the atomic particles decreases, the conditions in which classical mechanics are valid are eventually violated, and classical mechanics has to be replaced [as a tool for understanding] by quantum mechanics."

A protégé of Niels Bohr, and a man acknowledged as one of the great teachers of physics in the twentieth century, Landau was Kapitsa's closest colleague. An irascible perfectionist who liked to deny he had been a child prodigy even though he published a brilliant paper in quantum mechanics at the age of nineteen, Landau was arrested and imprisoned in the 1930s, charged with anti-Soviet activity and with being a Nazi spy, though he was Jewish. Kapitsa wrote directly to Stalin seeking Landau's release. "I beg you to give orders that his case should be very carefully considered"; Kapitsa acknowledged that his colleague's character was "bad," that he was "not easy to get on with [and] enjoys looking for mistakes in others [which] has made him many enemies," but denied that Landau could ever do anything seriously dishonest. When this did not produce results, he wrote to the foreign minister and then to the head of the secret police, demanding the release of Landau and personally guaranteeing that his colleague would "not engage in any counter-revolutionary activities." He also threatened that if Landau was not freed, he, Kapitsa, would resign. Landau later wrote that Kapitsa's activism on his behalf required "superb courage, great humanity, and crystalline integrity." Released from prison, Landau returned to his laboratory at the Institute for Physical Problems, where he shortly began to examine helium II and its strange antics in a new way.

Landau suggested that helium II be considered one large molecule, akin to a crystal. At absolute zero, Landau believed, helium II was 100 percent superfluid. As the temperature rose, "elementary excitations"appeared on the superfluid, particles known as phonons (quantized sound waves) and rotons (which move in exotic ways, such as in the reverse direction of their momentum). These particles, Landau guessed, comprised the normal fluid. From these ideas, Landau was able to formulate equations for the motion of the two fluids, normal and superfluid. He was also able to define viscosity as "the ability of a liquid to oppose movement" and the effective absence of viscosity as the inability to oppose movement. He reasoned that if anything were able to retard the flow of helium II into the capillary from the larger dewar in Kapitsa's experiment, the kinetic energy of the liquid would be reduced, its temperature would rise, and it would behave like (or become) helium I. But since the capillary walls did not impede the movement or change the energy level of the phonons, the fluid remained in the form of helium II, and only after heating did it exit the top of the capillary and turn the weathervane. According to Landau, the velocity of helium II as it penetrated into the capillary was low enough to permit the unimpeded flow through certain walls and against gravity. In other words, helium II was not faster than a bullet, as Kapitsa had contended; precisely the opposite was true. Helium II was the slowest yet the most persistently moving and unstoppable substance on Earth.

From the mid-1930s onward, scientists seemed to have most of the pieces of the deep-freeze puzzle spread out on a table in rough order, but they could not make the final assemblage. Along with the pieces having to do with magnetism, specific heat, and viscosity, there was another, provided also by Fritz London. He postulated that the difference between a metal in the normal state and in the super-conducting state had to do with an "energy gap" involving some-thing called the "Fermi surface," named after Italian physicist En-rico Fermi.* London believed that if one could figure out the micro-scopic mechanism of the energy gap that differentiated the two states, that would explain superconductivity and diamagnetism; he was also certain that the explanation would involve electron interac-tion at the Fermi surface.

Once London had identified the energy gap, Mendelssohn con-tends, the path to assembling the pieces was clear, but mapping that path was a "formidable task requiring a superb knowledge of elec-tronic phenomena in metals, great mastery of mathematical tech-nique, and, above all, brilliant but controlled imagination."

One man might not possess all of these talents, but a trio of men did: John Bardeen, Leon Cooper, and Robert Schrieffer. Their sum-mons to the puzzle board began in 1950, when a telephone call reawakened Bardeen's interest in superconductivity. In the fifteen years since London had directed attention to the energy gap, nuclear technology had been developed for the purpose of separating the

---

* The Fermi surface is not a real surface but a geometric description of the behavior of conducting electrons in a solid. A "surface" of constant energy, it separates energy states filled with certain kinds of electrons from those that are unfilled. At absolute zero, those particles with a certain spin would fill all the available energy levels up to the Fermi surface, but none above that. The absence of these particles is a characteristic of a substance in the superconducting state; in contrast, a substance in a normal conducting state has plenty of electrons of several kinds, which are the basis of normal resistivity.

isotopes of uranium and was now being turned to more benevolent uses. Looking into newly separated isotopes of mercury, two sets of researchers independently deduced an important mathematical relationship regarding superconductivity: the temperature at which a metal became superconducting varied inversely with the square root of its molecular weight. In 1950 F. Maxwell at the National Bureau of Standards and Bernard Serin at Yale University both arrived at the inverse-square-root formulation and prepared articles about it; Serin also telephoned his friend John Bardeen.

Bardeen was one of the prodigies of American science. The son of a medical-school dean and an artist, he graduated from high school at the age of fifteen and became one of the youngest students at Princeton's Institute for Advanced Studies in the mid-1930s. At Princeton and Harvard he studied the behavior of electrons in metals — and learned of the Londons' work in that area as it related to superconductivity. During the war Bardeen worked on magnetic fields given off by ships, and in the postwar era he teamed with Walter Brattain and William Shockley at Bell Labs to invent the transistor, for which the trio would win the Nobel Prize in 1956. Serin called Bardeen in 1950 because he knew Bardeen had come to believe that electron interaction at the Fermi surface of a metal could explain the onset of superconductivity,* and Serin's experiments and mathematical formulations provided additional evidence for this possibility.

Bardeen tried to shape from the various clues a comprehensive theory that solved all the superconductivity puzzles, but he could not do so for a few years. Then he was able to invite to his home base, the Illinois Institute for Advanced Studies, Leon Cooper, who had recently received his doctorate from Columbia University in quantum physics. Because not enough office space existed at the institute, grad students and postdoctoral fellows were crowded into

* A similar notion occurred independently and simultaneously to physicist Herbert Frölich while on a sabbatical year at Purdue University. Frölich's theory also predicted the isotope effect before it was discovered.

offices on Floor 3½ of a neighboring building, an enclave they labeled the Institute for Retarded Studies. In 1956 Bardeen asked Cooper to make room in his office for Robert Schrieffer, a doctoral candidate from the Massachusetts Institute of Technology (MIT). Bardeen had decreed that Schrieffer should spend a year in a laboratory as preparation for his doctorate, but when Schrieffer caused an explosion in the lab while welding metal in a hydrogen atmosphere, he was asked to concentrate on theory.

By then Bardeen had extended his guesswork beyond the energy gap described by Fritz London, adding the idea that superconductivity as a "phase transition" must be produced by a change involving the spin of the electrons. Phase transitions are the transformations that occur when a gas becomes a liquid or a liquid becomes a solid. In this case, the transition was from normal conductance to the superconducting state. While Cooper was traveling on a subway, the revelation came to him. Just above the temperature for the onset of superconductivity, electrons acted normally — they repelled one another, and one result was resistance to an electrical current. But when the temperature reached down to the transition there would be an interaction among electrons made possible by phonons. Two previously isolated electrons with oppositely directed spins would bond into a "Cooper pair." This bonding would be accompanied by some extra residual attraction that influenced all the other electrons, until none of them repelled one another. This was the superconducting state. Raising the temperature above the transition point would again break the Cooper pairs, making them repel one another, which produced electrical resistance.

The Cooper-pairs idea set the stage for a complete explanation of superconductivity. The trio now applied themselves to extending the Cooper-pairs idea to explain how such pairing could affect all the electrons at the Fermi surface. Working in part by analogy, they likened electron pairs of a potential superconductor to separated couples on a crowded dance floor. When a dance couple at the edge of the floor begins to move in unison, their motion affects the movements of other couples — or, in terms of electron pairs, it could be said that the first Cooper pair generates residual attraction that makes other pairs bond. On the dance floor, the added motion spreads from couple to couple in a wavelike pattern until all the couples on the floor are moving in unison. Another analogy: The Cooper pairs are like a heavy ball rolled on a mattress; the springs do not bounce back immediately as the ball passes, but stay down for a moment, during which a route is smoothed for other balls to precisely follow the first one's path — as though attracted by the first ball.

What was needed, as Schrieffer later put it, was a quantum-mechanical depiction of a wave function that choreographed the dance of 10^{23} couples — the number of electrons involved. This equation was so difficult to figure out that Schrieffer almost gave up on the task and considered changing his dissertation to one having nothing to do with superconductivity. But Bardeen convinced him to keep working at it for another month, while the advisor went to Stockholm to accept the 1956 Nobel Prize. During that time, Schrieffer figured out how to state the equation for the wave function of the dance, and when Bardeen returned, the trio tidied up the theory and set to work trying to see if it would explain all previous experimental results.

As they did, their excitement grew, because what came to be called — after their initials — the B-C-S theory could explain the abrupt onset of superconductivity, the way in which magnetism was expelled from the interior of a solid, the rise in the specific heat at the transition temperature, and even superfluidity. All of these phenomena had to do with the action of Cooper pairs. Above the critical temperature, the pairs were apart or broken, and resistance and other normal attributes, such as magnetic susceptibility, existed; at the critical temperature, the Cooper pairs came together, a bonding that drastically changed the structure, permitting electric currents to course without resistance, eliminating interior magnetism, raising the heat capacity, and enabling liquid helium to flow endlessly. The connection between superconductivity and superfluidity now became clear: the chief characteristic of both was the ability to sustain currents of electricity or of helium atoms at a constant velocity for long periods of time, without any apparent driving force.

Bardeen recalled that in early 1957, when he, Cooper, and Schrieffer compared their theory with other scientists' experimental data, "we were continually amazed at the excellent agreement obtained." When they came across apparent discrepancies, they discovered by redoing the math of the other experiments that these calculations contained errors, not contradictions of their theory. They submitted a short communication to the *Physical Review* and followed it up several months later with a more detailed paper. Superconductivity, the greatest puzzle in the vicinity of Ultima Thule, had at last been solved.

Mastery of the Cold

FOR TEN THOUSAND YEARS, civilization had progressed by means of ever greater control of heat; in the twentieth century, progress came about through control of the cold.

In 1912 Clarence Birdseye went ice fishing in Labrador, in temperatures 20 to 40 degrees below freezing. Imitating the natives, after hauling up his catch through ice several feet thick, Birdseye would throw it on a line over his shoulder. In minutes, because of the frigid air temperature, the fish would freeze solid. He noted that even when these fish were cooked weeks later, they tasted fresh, unlike fish frozen by slower methods. A chemist, Birdseye knew the general rule that the longer it takes for crystallization to occur, the larger the individual crystals that are formed. The reason for the fresher taste of quick-frozen fish, he determined through research, was that while in regular freezing ice crystals formed and grew in the frozen flesh, rupturing cell walls and compromising cell integrity, in quick-freezing solidification took place so rapidly that only very small ice crystals formed, and the cell walls did not break. Maximum crystal formation occurred between $-1°$ and $-5°C$; by passing rapidly through that zone on the way to lower temperatures, quick-freezing avoided excess crystallization.

It took Birdseye another ten years to work out a commercial fast-freezing technique, using a calcium chloride brine solution that kept the temperature at $-45°F$ and applying the intense cold to the

undersides of metal plates along which the fish traveled in a con-
veyor-belt mechanism. Achieving the low temperature was the least
difficult problem; Birdseye's operation would have been possible
fifty years earlier had anyone realized the commercial potential of
quick-freezing food.

Birdseye raised some capital and built his first plant in New York
in 1923; its processing machine weighed 20 tons, making it all but
immovable. With additional capital and investors he built a second
plant in Gloucester, Massachusetts, in 1924, and he was soon doing a
brisk business and examining the possibility of freezing foods other
than fish. By 1928, 1 million pounds of quick-frozen foods were
being sold annually in the United States, the preponderance of them
by Clarence Birdseye's company; just before the 1929 stock market
crash, the Birds Eye company was bought by Postum cereals, the
precursor of General Foods.

Fast-freezing machinery had grown progressively lighter in
weight, to the point where it would shortly be movable from site to
site to take advantage of peak harvests. Food experts waxed rhap-
sodic that this would allow utilization of ever greater percentages of
harvested fruits and vegetables, leaving still less to rot in the fields or
in storage. The main obstacle to wide acceptance of frozen foods was
the consumer, who looked askance at such items. Still, the promise
for the future was palpable: in the words of a contemporary industry
journalist, "Strawberries could be eaten in December, and you didn't
have to shell peas or wash spinach any more." A test of the retail-
market appeal of fresh-frozen fish, vegetables, and fruits was held in
1930, but the arrival of the Depression soon put Postum's plans for
an enormous expansion of the industry on hold.

The first third of the century also saw the emergence of two other
industries based on mastery of the cold: artificial refrigeration and
air conditioning. At the turn of the century, though iceboxes were
in regular use in about half of American homes, only 1 percent of
U.S. households featured bulky, expensive, artificial-refrigeration
machines. These were variations of what Linde had produced for

industrial firms, based on the laws of classical thermodynamics and advances in compression machinery. Further spread of mechanical refrigerators was hampered by the need to connect the machines to local sewer systems to ensure frequent disposal of their dangerous refrigerant chemicals — ammonia, ethyl chloride, or sulfur dioxide. A less dangerous refrigerant, carbon dioxide, was not as widely used because it required more pressure to function properly. By 1915, using knowledge about refrigerant chemicals and low-temperature handling techniques pioneered by basic researchers such as Dewar and Onnes, the General Electric Company developed better carbon-dioxide-based refrigerators that were independent of sewer connections, self-contained, and small enough for the average home; during the following fifteen years, a million such refrigerators were sold by GE and its main competitors, Kelvinator and Frigidaire.

In 1902 the new building of the New York Stock Exchange opened, with an air-cooling system for its trading room, designed by Alfred Wolff. In 1905 Stuart Cramer coined the term *air conditioning* to describe what the machinery he was installing would do to the air in a southern textile factory — control its volume, humidity, temperature, and recirculation, with the emphasis on humidity control rather than on interior cooling. The third father of the field was Willis Carrier, a young engineer who fitted factories with temperature- and humidity-control units for the Wendt Brothers of Buffalo between 1908 and 1915, when he and a half-dozen compatriots were fired because the likelihood of American entry into the war convinced the Wendts that the future for air-control units was dim. Carrier and the engineers formed their own firm, which by 1918 had completed installations in factories serving more than seventy industries. The units they installed were huge, central air-conditioning machines, too large for use in private homes.

The United States emerged from the Great War in a stronger economic position than any other combatant nation; this economic strength began to express itself in many ways, among them growing use of the cold, for instance in skyscrapers and in motion-picture theaters. As commercial skyscrapers reached ever higher into the sky, their architects realized that at about twenty stories above the street, windows should not be opened, to prevent the havoc of wind gusts, more prevalent there than at ground level. Recognizing also that in summer the absence of outside airflow might make the buildings unacceptably hot, developers began to install central air conditioning in their taller buildings and to permanently seal their windows. In the 1920s, air cooling was installed in American motion-picture theaters, making them for the first time usable in summer and creating widespread appreciation among patrons for cool interior air. That appreciation, combined with the growing popularity of home refrigerators, led in 1928 to the Carrier firm's designing the first single-room home air conditioner. Because the Carrier machine controlled humidity as well as temperature — an exceedingly difficult task — the manufacturer was unable to price it lower than $3,000, which meant that these first air conditioners were beyond the reach of almost all Americans.

The Tennessee Valley Authority and other rural-electrification projects of the 1930s spread the use of artificial refrigeration into new geographic areas. Shortly, despite the Depression, the majority of American homes featured refrigerators. Only the eclectic continued to use true iceboxes in their homes when they could afford the mechanical alternative: the poet e. e. cummings refused to buy a refrigerator and give up his icebox, because he was fond of the man who twice a week delivered ice to his Greenwich Village flat.

Tracer bullets had easily exploded the World War I dirigibles that had been filled with hydrogen gas rather than helium, because helium had been too expensive and difficult to obtain. After the war, as a byproduct of examining routes to improving the efficiency of extracting and manufacturing gaseous helium for dirigibles, the U.S. Bureau of Mines in 1924 first produced liquefied natural gas (LNG), a combination of methane and other gases that liquefaction could reduce in volume 580 times. Here was a fourth industry whose existence depended largely on the ability furnished by application of the cold to condense its products for transport and storage. Widespread use of LNG as a heating fuel was hampered by the relative cheapness of fossil fuel oil found at the same locations as the gas and by a disastrous explosion of a natural-gas production facility in 1944. These factors put the nascent industry on hold for several decades, until the 1970s, when the Organization of the Petroleum Exporting Countries (OPEC) hiked oil prices to the point where LNG became an attractive alternative and supertankers were built to make its transport equally economic.

Large-scale production of liquefied gases for other uses started before that of LNG, because the numbers were so compelling: a ton of gaseous oxygen occupied 26,540 cubic feet at normal temperature but only 31 cubic feet when liquefied; the ratios were similar for nitrogen, hydrogen, argon, neon, and helium. All sorts of manufacturers began to rely on the liquefied, deeply cooled products of air-separation plants that could be easily transported and then returned to gaseous form when needed. The technology of electric lighting depended on the availability of inert gases such as neon and argon to fill light bulbs. Steel making could no longer be accomplished properly without blast furnaces into which compressed oxygen was introduced. For arc welding — as predicted by Georges Claude, the developer of the French cryogenic industry — oxygen now supplanted acetylene. Liquefied gases were also used in the production of many industrial chemicals, fertilizers, and photographic films.

Leo Dana, the American chemist who had spent a year working with Kamerlingh Onnes, came home and found a job with the Union Carbide Company in 1923, shortly after it had bought the Linde air-separation operation in the United States. Among the first problems Dana tackled was the need for better insulation, so that liquefied gases could be more readily transported and stored. His "powder in vacuum" insulation boosted Union Carbide's ability to sell its products, and it enabled basic-research laboratories to maintain a steady supply of liquid gases.

Too small to be noticed as a market just then was the use of liquefied gases as fuel for the experimental rockets being constructed by Robert H. Goddard in the United States and a small group of scientists in Hitler's Germany. The rocketeers needed the big thrust that controlled burning of hydrogen and oxygen could provide, and only by using liquefied gases could they pack enough fuel into their rockets to enable the devices to reach thousands and then tens of thousands of feet up into the sky.

World War II was a turning point for all four cold-based industries. In 1940 the U.S. Army wanted a portable oxygen separator for use in field hospitals; when Union Carbide refused to sell its equipment to the government, the military turned to William Giauque of Berkeley, whose ingenuity had already produced the demagnetization process for reaching close to absolute zero, and who now advanced the utility of liquefied gases by making a new, mobile oxygen separator for the Army, later adapted for civilian hospitals.

Days after the Japanese attacked Pearl Harbor, in December 1941, meatpackers in Chicago boned, quick-froze, and shipped to the Pacific a million pounds of beef, chicken, and pork. This demonstration intensified the military's determination to feed frozen foods to personnel in faraway theaters of combat. The sheer quantity of food required by the military created a greater need to avoid wasting crops or livestock. Moreover, because so much fresh food went to the military, there was higher civilian demand for frozen fruits, vegetables, meats, and fish. A last and unexpected spur to the frozen-foods industry was also due to the war: since many automobile showrooms stood empty because they lacked product, their owners sought other uses for their space, and a significant number of the showrooms became frozen-food distribution centers.

Enlarged military bases in areas that had previously been sparsely populated, such as the deserts of the Southwest and the near-tropical areas of the Southeast, started to attract civilian populations to serve them, increasing the demand for air conditioning. In the dry regions, the technique of "evaporative cooling" became fashionable, partly because it was cheap and easy to operate: a fan blew dampened air through a home, and the dry air in the house absorbed the humidity, cooling the rooms. The availability of evaporative cooling made cities such as Phoenix more habitable, and their populations grew rapidly. Humid areas such as Florida required more expensive air-cooling techniques; despite the cost, new migrants to the Southeast gravitated to developments featuring homes with central air conditioning already installed, as life in the South began to seem unimaginable without artificially cooled air.

The explosive growth of America's suburbs in the postwar era brought with it spectacular increases in the use of frozen foods, refrigeration, air conditioning, and liquified gases. A spacious, air-conditioned supermarket selling a wide variety of frozen foods helped attract people to the suburbs; the annual consumption of frozen foods leapt toward 50 pounds per person, with two hundred new products introduced each year. Although only one out of every eight homes in the United States had air conditioning in the 1960s, four out of ten in the Sunbelt region had it. Nearly every American home featured a refrigerator, and many newer homes, more than one. Greater use of the cold became inextricably associated with America's advancing standard of living — an index of material comfort that expresses the degree to which people control their environment. And as the American standard of living rose, refrigeration, air conditioning, frozen foods, and other products made with cold technology no longer were luxuries but were judged necessities of modern life.

Meanwhile, what had once taken Onnes years, cumbersome equipment, and considerable expense to produce — a few liters of liquefied helium — by the 1950s could be done routinely in almost any laboratory, after the invention by Sam Collins of MIT of a liquefier no bulkier than a home appliance. Shortly, the technique was extended to commercial manufacture of liquid helium. New or improved uses for many cold-liquefied gases emerged: the employment of liquid nitrogen to store blood and semen, the use of liquid hydrogen and liquid oxygen in the rockets of space-exploration programs, liquefied-gas coolants and scrubbers for nuclear reactors, liquid-helium traps for interstellar particles. Artificial insemination of dairy herds became more widespread as the ability to store semen advanced. Miniature Linde-Hampson systems with Joule-Thomson expansion nozzles made possible more portable liquid-cooled infrared sensors; these systems were utilized mostly by the military — for "smart" missiles, projectiles, and night-vision systems — but they were also useful in detecting fires and, in medicine, for detecting diseased tissue. Surgical technique was advanced by the introduction of cryosurgery; a probe carrying liquid nitrogen into the brain or the prostate could do what a metallic knife could not: first tentatively cool a section of tissue, allowing the surgeon the latitude of evaluating the probable results of surgery before deciding whether to permanently destroy the target area.

As space shots reached beyond Earth's atmosphere, boosted into orbit by combinations of hydrogen and oxygen fuel stored in ultra-cold liquid form in the immense rockets, scientists received the first experimental verification that the temperature of interstellar space was within a few degrees Kelvin of absolute zero. Shortly, liquid oxygen and liquid hydrogen provided the fuel to send men to the moon and return them to Earth, and cold-control apparatus permitted them to stay alive on the journey.

In the universe described by Newtonian physics, nothing smaller than a cannonball shot at a solid wall with great force could push through that wall; the force and mass of the cannonball overcame the energy of the atoms within the wall, which in all other cases was great enough to repulse the energy of the atoms attempting to penetrate it. In the universe described by quantum physics, sub-atomic particles can also sometimes pass such a barrier, by a process known as tunneling, in which the particles do not overcome the energy of the atoms in their way but instead find a route between the atoms of the wall. In the wake of the revelation of the B-C-S theory in 1957, two scientists half a world apart made discoveries about subatomic tunneling related to superconductivity. In Tokyo, a graduate student in physics employed by the Sony Corporation, Leo Esaki, described tunneling effects in semiconductors at low temperatures, and made what are now called Esaki diodes. In Schenectady, New York, the Norwegian-born graduate student Ivar Giaever was working for General Electric and taking a course at Rensselaer Polytech when he realized that the tunneling of electrons might be used to measure the energy-band gap long ago identified by Fritz London as existing at the Fermi surface in a superconductor. On April 22, 1960, he made a metal sandwich, a layer of insulation between two thin plates of metals; when the outside layers were in the normal state, electrons in a current could tunnel through the insulation, but when one of the layers was in the superconducting state, no tunneling occurred. The effect was measurable, but it had not yet been explained.

A third graduate student, Brian Josephson, at Cambridge University in 1962, drew on the insights of Esaki and Giaever and on Cooper's work with electron pairs — the linchpin of the B-C-S theory — to predict that Cooper pairs could tunnel through the metal sandwich, even when both outer layers were superconductors. From this theory he drew the implication that if the voltage in the current coursing through the layers fell below a certain level, or if a magnetic field disturbed the superconductivity, the Cooper pairs would break, and that the change in the voltage would be sudden and measurable. Josephson's teacher in the Cambridge course, Philip Anderson, took this idea back to Bell Labs and had the equipment built to prove it experimentally. Several companies began production of "Josephson junctions" to record or measure minute changes in electrical voltage and magnetic fields. These junctions became the heart of new gadgets known as SQUIDs (superconducting quantum interference devices). SQUIDs are used in voltmeters for low-temperature experiments, including those on space satellites; in magnetometers sensitive enough to pick up the magnetic field of a passing submarine, a human brain, or a heart, or even of a single neuron; and in making speedy logic elements and memory cells in computers. An experimental Josephson junction computer was constructed in the 1980s, with parts of the works sitting in special ultracool liquids, a machine 100 times faster than the usual computers. Indicating the importance that scientists attached to the ideas of Esaki, Giaever, and Josephson, they were jointly awarded the Nobel Prize for 1973.

Josephson junction innovations did not garner much public attention, which went instead to the yearly, breathtaking improvements in microchips and semiconductors. Nor did the public realize that as electronic technologies advanced, their need for cooling became greater. Supersonic-aircraft flight speeds had brought to the fore the problems of cooling the electronic gadgetry necessary to operate them, to reduce the size of the devices and improve their reliability. Phalanxes of engineers began devoting themselves to a new specialty, the technology of removing heat from all sorts of electronic equipment. Back in 1947, the first chips contained one component each; by the 1970s, a chip could hold close to 100,000 components, including transistors, diodes, resistors, and capacitors, and there was a need to cool such chips while they were operating and also during their manufacture, often to cryogenic levels. According to a textbook, the "dramatic" increase in miniaturization mandated that "thermal considerations . . . must be introduced at the earliest stages of design." The more complex the electronic gadget — the computer, the television set, the mobile telephone — the more likely it was to contain parts manufactured at temperatures hundreds of degrees below freezing. Cooling electronic machinery during manufacture and performance became the fifth large cold-based industry of the modern era.

The need for cooling renewed interest in thermoelectricity, the effect discovered by Peltier in 1834 and more fully explored by Kelvin in 1854, in which cold can be generated by electric currents flowing across two conductors made of different materials. Twentieth-century research revealed that the thermoelectric powers of semiconductors are much greater than those of metals and that thermoelectric capacities can be expanded at low temperatures (80 to 160 K) by the application of a magnetic field. New semiconducting materials made from compounds never known before the last few decades are now used to create very small refrigerating devices for computer components and other sensitive electronic circuitry, including miniature lasers.

The ultimate point in miniaturization and cooling may have been reached by the Baykov Institute of Metallurgy in Moscow and the Odessa Institute of Refrigeration, which in the 1990s created thermoelectric coolers on single crystals made from solid solutions of bismuth and antimony compounds. Single-crystal coolers are being used experimentally for infrared detectors, light-emitting diodes and lasers, and devices for night viewing, astronomical observations, ground and space optic communication, missile guidance, and target illumination.

In January 1962 Lev Landau was involved in a car crash in Russia that broke nearly every bone in his body and put him into a coma for fifty-seven days; perhaps goaded by his being near death, the Nobel committee awarded him the Nobel Prize in physics in December 1962 for his decades-old work on the theory of condensed matter and superfluidity. Soviet scientists were chagrined that Landau would be honored without the prize being jointly awarded to Pyotr Kapitsa, who had done the basic experimental work on superfluidity, but they hoped Kapitsa's turn would come. It took another sixteen years, until 1978, when Kapitsa shared the Nobel Prize with two American physicists who discovered cosmic microwave background radiation. Landau never regained his health or returned to his laboratory work before his death in 1968. By then, interest in superfluidity had surged again, especially as magnetic cooling enabled investigators to lower temperatures into the "millikelvin" range of a few thousandths of a degree above absolute zero.

In the 1970s, demagnetization was alternated with other cooling techniques, in a process that deliberately mimicked the operation of the ideal engine Carnot had imagined, to bring down temperatures and to produce a continuing series of revelations. Physicists had long assumed that superfluidity might be a widespread phenomenon, not simply a property of one form of liquid helium at low temperatures. These guesses received some verification from the work done in 1971 by three physicists at Cornell University, Robert C. Richardson, David M. Lee, and graduate student Douglas Osheroff. They discovered that at a temperature of two-thousandths of a degree above absolute zero, a new form of the element, known as helium-3, a form that had previously been thought capable of becoming a superfluid but had proved elusive, could be made into a superfluid. Their results were so unexpected that one prestigious journal initially rejected their article about the discovery. Superfluid helium-3, they found, was "anisotropic" — like a crystal, it appeared to have different values for properties depending on which axis was being measured. The new superfluid could act like a quantum microscope, permitting the direct observation of the effects caused by interactions among atoms. Richardson, Lee, and Osheroff would win the 1996 Nobel Prize for their work on creating a superfluid from helium-3.

Part of the reason they won the prize was that shortly after their discovery, using the Cornell trio's data, astronomers came to believe that the transition of helium-3 into a superfluid was analogous to the formation of the vast structures in space called cosmic strings, in the microseconds after the "big bang," and that superfluidity as a state or attribute of matter might be present in rotating neutron stars, thousands of light-years distant from Earth. This realization was one instance of a grand coming-together of branches of science that had once seemed separate and unrelated. Connections were found that bolstered the links between the study of the behavior of matter at ultra-low temperatures, the study of subatomic particles, and the study of the origins of the universe. The ability to control the ultracold was the key to all three. Supersensitive devices based on cold technologies had become capable of measuring the entire electromagnetic spectrum, of registering images of the radiant heat of celestial objects in the infrared, millimeter-wave, and microwave range, as well as images of gravity and magnetic emanations. These could be used to identify various relics from the early days of the universe, leftovers from the era of the big bang. Among the dozen types of emissions scientists tried to find in the sky were fractional electric charges such as quarks; one theory holds that the universe began as a sea of quarks, some of which could have survived the big bang. Other emissions include WIMPs, (weakly interacting massive particles), and "stochastic" or random gravitational radiation, which might give clues to what happened during an early "phase transition" of the universe. Helium-3 was also found in the residues from volcanic eruptions, sometimes encapsulated or trapped between carbon atoms, and identified as remnants from the time of the formation of the earth.

In professional conferences with such titles as "Inner Space/Outer Space," physicists explored the efforts to record the big-bang relic particles of deep space and the efforts being made to understand subatomic particles in earthbound laboratories. Since World War II, physicists had relied for investigation of these latter particles on linear accelerators that raised the particles' speed to several thousand miles per hour and let them smash into obstacles, or each other, and disintegrate into interesting pieces.

It had long been known that a magnet's force could be amplified by means of electric current coursing through wire wrapped around it. When it became possible to make wires that were superconducting, and wrap them around magnets, the resulting current raised the power of the magnets even more.* Such superconducting magnets were applied to increase the sensitivity and effectiveness of magnetic resonance imaging (MRI) devices, employed in medicine to detect diseases such as cancer in soft tissues that x-rays could not reveal. They also became critical components in masers, microwave precursors of lasers that were used for communications and are still used to detect remote astronomical events. In high-energy physics, superconducting magnets were incorporated into the design of new linear accelerators, to raise the speed of the beams of subatomic particles being hurled at one another. Important discoveries about subatomic particles were made with these new accelerators, such as the Tevatron at the Fermi National Accelerator Laboratory in Illinois.† Superconducting magnet colliders showed such promise that governments in the United States and in Europe agreed to build multibillion-dollar "superconducting supercolliders" (SSCs); one expert calculated that to raise particles to the same energy level as the American SSC would make possible, an accelerator constructed in the same way as the 2-mile-long one at Stanford University would need to be 100,000 light-years long.

Just after completion of the first several miles of the 54-mile tunnel near Waxahachie, Texas, that would hold the American SSC, and a test of the first of the six accelerator stages in which hydrogen ions were successfully smashed, in the early 1990s congressional budget cutters shut down the SSC, arguing that $2 billion had already been spent with little to show for it. Following the cancellation

* While the application of a magnetic field to a superconducting material could make that material lose its superconductivity, when an insulated superconductor was wrapped around a magnet, amplifying the magnet's power, the superconductor was not affected adversely by the magnet.

† Perhaps the most important of these was the 1995 discovery of the "top quark." Decades earlier theorists had predicted six different kinds of quarks. Five "flavors" had been found and identified in the 1970s: the pairs up and down, strangeness and charm, and the single one called bottom. Thus the last to be discovered was the other half of the third pair, called the top quark.

of the SSC, several hundred scientists and engineers left the United States to join the European project; with this shift, the center of research on particle physics relocated to near Geneva, site of the Large Hadron Collider.

In 1975 physicists suggested a new way to study atoms, not by accelerating them but by simultaneously slowing and cooling them. In 1985 Steven Chu of Bell Labs ingeniously used lasers fired from six directions to make what he called an "optical molasses," a "trap" to intercept and slow incoming atoms to speeds measured in inches per second. The trap confined a few thousand atoms to one spot, at a temperature of 240-millionths of a degree above absolute zero. Chu's concept was extended by an even better magnetic-trapping device perfected in the following years by William D. Phillips of the U.S. National Institute of Standards and Technology, then explained theoretically by Phillips, Chu, and Claude Cohen-Tannoudji of the École Normale Supérieure in Paris, who cooled the atoms even further, to within one-millionth of a degree above absolute zero. Chu, Phillips, and Cohen-Tannoudji were awarded the 1997 Nobel Prize for their achievement.

While this exciting development was invigorating research into sub-atomic particles, an astonishing advance was made in the field of superconductivity. Since the time of Onnes, scientists had been trying to find materials that became superconductive at temperatures higher than that of liquid helium, believing that only when super-conductivity could be produced easily and economically could it be put to the tremendous practical uses Onnes and every subsequent thinker in the field had envisioned — free-flowing electric currents, more powerful magnets, a world with a near-infinite capacity to conserve and distribute its energy supply. Onnes had discovered the onset of superconductivity in mercury at 4.19 K; in the ensuing seventy years, the record for the "critical temperature" at which superconductivity commenced had been raised only 19 degrees, to 23 K for niobium, a rare metal. At IBM's Zurich laboratory, in the

early 1980s Karl Alex Müller and his junior associate Johann Georg
Bednorz began working with metallic oxides to see if they could be
made superconductive; oxides — combinations of oxygen with
other elements — were a bit of an odd choice, since some were used
as insulators, and many had no electrical conductivity at all, although a few had been shown capable of becoming superconductors. Working more like chemists than physicists, Bednorz and
Müller mixed compounds, baked the mixtures in ovens, and then
chilled them to liquid-hydrogen temperatures. They did their work
without lab assistants, and without much encouragement from their
colleagues in the laboratory; Bednorz even had to steal time from his
regular assignments to perform the oxide experiments. After two
and a half years of trying various compounds, on January 27, 1986,
they succeeded in making an oxide of barium, lathanum, copper,
and oxygen that became superconducting at 35 K. Being cautious,
telling almost no one of their accomplishment, they prepared an
article in April for publication in September 1986 in *Zeitschrift für
Physik*.

An explosion of research and excitement followed that publication, as laboratories in Japan, China, England, Switzerland, and the
United States jumped into the chase for a compound that would
become superconducting above 77 K, the temperature at which nitrogen liquefied. Since the liquefaction of nitrogen had become routine and inexpensive, if liquid nitrogen could be used to produce
superconductivity, scientists reasoned, there ought to be no limit to
employing superconductivity for the profit of whatever entity could
patent the compound having the highest "critical temperature." A
frantic six-month scramble among the laboratories led to the jam-
packed "Woodstock of Physics" meeting in New York City on March
18, 1987, at which the major groups reported their recent research —
some accomplishments so new that the ink was not yet dry on
articles about them. The winner of the chase for the compound that
was the easiest to create, and that had the highest critical tempera-

ture, was Paul Chu's laboratory at the University of Houston, whose "1–2–3" compound became superconductive at an astounding 93 K.

A media frenzy followed, reaching and involving the uppermost levels of governments on several continents, as superconductivity achieved at liquid-nitrogen temperatures (above 77 K) was touted as the key to everything from Star Wars missile defense systems to superfast computers to energy storage and transmission devices that would drastically lower the price of electric power. In the delirium over what seemed the ultimate use of the extreme cold, an eighth-grade science teacher — the daughter of an IBM physicist — used the "Shake 'n Bake" method to cook a wafer of the new compound in a regular oven, then placed it in a dish of liquid nitrogen and magically floated above the dish a tiny magnet. Cornelis Drebbel would have been pleased.

Müller and Bednorz were awarded the 1987 Nobel Prize in physics, and there were high hopes that the 1990s would be the decade in which "high-temperature superconductivity" (HTS) would revolutionize the world. When these overblown expectations were disappointed by the difficulty of fashioning the new compounds into electrically conductive wires, and of constructing ways to maintain them at liquid-nitrogen temperatures, it seemed as though a balloon had burst. However, sure and steady progress in utilization was made in the decade after the 1987 Woodstock of Physics — an industry journal claimed that the "pace of utilization" of HTS was on a par with that achieved by other high-tech "overnight sensations" such as microprocessor chips. More than one hundred new HTS compounds have been created, with onset temperatures as high as 134 K, nearly twice as warm as liquid nitrogen.

Equally important, the practical use of superconducting wires has begun. In Geneva, the public utility now has a transformer wound with HTS-compound superconducting wires to step down the voltage from the country's power grid; since the new transformer runs without oil, the likelihood of the fires and pollution that often occur with regular transformers is drastically decreased. In Detroit, a contract has been awarded to the American Superconductor Corporation to produce a 400-foot-long superconducting line for that city's public utility by the year 2000; the line's 250 pounds of superconducting wire will carry as much current as 18,000 pounds of the existing copper wire. Among the costs saved are significant environmental ones, since the creation of 250 pounds of superconducting wire uses up considerably fewer natural resources than does the extraction, refining, and manufacture of 18,000 pounds of copper wire.

In North Carolina, the public utility offers superconducting magnetic-storage devices to commercial customers for use in countering unexpected power surges and dips. Other projects nationally include a superconducting generator coil, a 125-horsepower motor, many times smaller than usual motors, and improved SMES (superconducting magnetic energy storage) systems, in which the magnets are charged during off-peak hours when demand is low, enabling them to make more power available to the grid when demand rises. The U.S. Department of Energy estimates that if all public utilities switched to superconducting transmission and distribution lines, they would be 50 percent more efficient. Expected for the future is a shift from generating plants in and near cities to ones in remote locations, or ones that use less expensive solar or geothermal energy sources to produce electric power, which can then be cheaply and efficiently transmitted for use in population centers. An added bonus expected from these sources is that more efficient transmission of power will diminish pollution.

The new HTS superconductors show equal promise for electronic equipment: they are being used to filter signals from noise in cellular-phone base stations, improving cell-phone reception; they are reducing imaging time, improving resolution, and lowering the costs of MRIs. Perhaps the most unexpected use is in sewer and water-purification systems: iron compounds are salted into the liquids, where they bond with undesirable bacteria and viruses, forming substances that superconducting magnets can then attract and remove. A similar "magnetic separation" process is being used in portable devices to clear contaminated soil sites, including sites that contain the radioactive compounds called actinides. Other high-tech applications are on the drawing board. Superconductivity applications seem likely to become the sixth major industry based on mastery of the cold.

In terms of numbers of people and industries served the technologies of cold are at an all-time high. Virtually all American homes have refrigerators, and most have air conditioners, as do all modern business buildings from factories to warehouses to corporate headquarters. In the Far East, the major source of energy for electric power is rapidly becoming LNG. Oxygen transported in liquefied form is in use in all hospitals. Other liquefied gases are critical to dozens of manufacturing processes; worldwide, annual sales of such gases total $10 billion, with the American-based company Praxair accounting for about half of that. Throughout the industrialized countries, most people daily use electronic devices made with, food preserved by, or chemicals manufactured by means of liquefied gases and the cold they produce. Increasingly, what separates the "have" from the "have-not" nations of the world is that the first group makes more use of the cold.

Still more uses are coming. In 1998 two milestones were reached in the use of superconducting magnets. In one, the first 18.4 kilometers (11.4 miles) of track for Japan's "maglev" — magnetic levitation — train was opened. The sets of magnets in the train, the tracks, and the ancillary equipment float the train millimeters above the guide track and serve to accelerate it along the track at speeds much faster than can be achieved by any system in which the train and track are in contact. In the second milestone, superconducting magnets were employed in brain surgery; in St. Louis, in December 1998, the magnets were used to direct a surgical instrument around corners and on a curved path through the brain, avoiding vital sections, to perform a tumor biopsy. The process was considerably less invasive than conventional methods and is expected to be used in the near future to treat motor disorders such as Parkinson's disease that are centered in the brain, as well as to treat cancer and heart disease in other areas of the body.

The new Kamerlingh Onnes Laboratory at Leiden opened in late 1998, and while it, too, works with HTS materials, its main mandate is to explore many frontiers of physics at temperatures capable of being generated by liquid helium. At one end of the research "factory," an automated production facility manufactures large quantities of liquid helium, which is then held in 5,000-liter containers and siphoned off into smaller ones. A screen saver on the computer in the production facility asks, as Onnes might have done, "How much helium have you wasted today?" Containers of the precious fluid are wheeled down corridors to the other end of the building, where a dozen cryostats and associated measuring equipment are located on concrete-and-steel platforms specially constructed to eliminate all vibrations.

Scientists working at the cutting edge of physics increasingly choose to investigate all sorts of physical phenomena by means of low temperatures, says the Leiden lab's current director, Jos de Jongh, because at micro- and millikelvin temperatures "you can eliminate all the extraneous influences," such as radiation and vibration, and be more certain that the only variable is the phenomenon under study. For instance, de Jongh, his graduate students, and visiting researchers from several countries recently completed studies of how large a metallic cluster must be before it stops behaving like a collection of atoms and starts behaving like a bulk object; the answer was in the range of 150 atoms. Other studies in the lab have used special cameras operating at ultra-low temperatures to observe the crystallization of helium-3 at millikelvin levels.

The use of ultra-low temperatures in basic research on the structure of matter entered a new phase in early June 1995, when a team of physicists at a research coalition of the National Institute of Standards and Technology and the University of Colorado, led by

Carl E. Wieman and Eric A. Cornell, had an experience paralleling those of Faraday in 1823, Cailletet in 1877, and Onnes in 1911. While conducting an experiment at the lowest temperatures they could reach, they produced a blob of material never seen before. In this instance, the blob was what had eluded scientists for seventy years, a Bose-Einstein condensate (BEC) — that gas of atoms whose existence had been postulated by Einstein in the 1920s but that had not been definitively known to exist until 1995. The temperature was 170-billionths of a degree above absolute zero, and the blob was not even seen directly, since it was immediately destroyed by a laser probe flashing through it, but its image remained on a computer screen. Further experiment reduced the temperature to 2-billionths of a degree Kelvin, more than a million times colder than interstellar space.

This was a highly significant event — combining the production of a form of matter that no one before this had been certain could be generated on Earth with the reaching of the deepest and most profound cold, a cold almost beyond imagining.

Only weeks later, another BEC was created at Rice University, and within months, others were brought into existence at Stanford and at MIT. Physicists were elated over the possibilities of using BECs to study the mechanisms behind superconductivity and superfluidity — superfluid helium was believed to have some of the characteristics of a BEC — and to examine other aspects of atoms and elementary particles. "If you want to speculate wildly," Cornell told a reporter, "you could imagine an atomic beam analogous to a laser beam — one that could move or deposit single atoms to build a molecule-size structure."

That wild conjecture became reality in less than two years, in January 1997, when a team headed by Wolfgang Ketterle at MIT created an "atom laser" from a BEC. In 1998 Ketterle's group was able to magnetically manipulate a BEC-based atomic laser to do what Cornell had predicted: move atoms around, and form complex molecules. It was an indication that whatever could be postulated in terms of subatomic particles at ultracold temperatures might well be realized in the near future.

In February 1999 a team at Harvard headed by Lene Vestergaard Hau used the BEC and laser-cooling techniques to produce an environment only 50-billionths of a degree above absolute zero and to slow the speed of light to a mere 38 miles per hour. As with Ketterle's atomic laser, no immediate applications were expected from the Hau group's feat, but it was believed that in ten years' time, many practical uses might be developed. In mid-June of 1999, as this book was going to press, the MIT group announced another breakthrough. For the first time, the scientists had quantitatively measured zero-point motion in a BEC that, according to Ketterle, "has no entropy and behaves like matter at absolute zero."

As the research on subatomic particles goes forth in the coldest temperatures imaginable, opening up possibilities for studying and manipulating the subatomic building blocks of matter, so do projects in other scientific fields that rely on mastery of the cold, among them some of the most advanced projects now being conducted. To study the farthest reaches of the universe from Earth, electronic detectors chilled by liquid helium have recently been set up near the South Pole, in an ambitious astronomical project, AST/RO, the Antarctic Submillimeter Telescope and Remote Observatory. And to study asteroids and other solid bodies in deep space, an unmanned probe has been launched, powered by a new propulsion system inspired by science fiction, an engine that moves the ship through space by means of squeezing the energy out of the ions of a rare gas, xenon — a gas obtained through air-separation processes, and maintained on board the spacecraft as an ultracold liquid.

If the present direction and volume of research are any guide, a large proportion of tomorrow's technological advances, and of tomorrow's discoveries about the composition of matter and the nature of the universe, will be made in the vicinity of absolute zero and will be based on our mastery and manipulation of the cold.

Acknowledgments

Notes

Index

Acknowledgments

I wish to thank the Alfred P. Sloan Foundation for a generous grant enabling me to complete this book. The Writers Room in New York City afforded me shelter through this project, as it has through others; my colleagues and the staff there have been a source of unflagging support and encouragement. Much of the research was conducted at the main branch of the New York Public Library at 42nd Street and in the recently opened Science, Industry and Business Library (SIBL) at 34th Street; the library also provided me with facilities in its Wertheim Room. Other American libraries consulted include those at Columbia and New York universities, the Library of Congress, the Scoville Library in Salisbury, Connecticut, and the Alfred H. Lane Library at The Writers Room.

In London, I worked extensively in the unrivaled historical collections of the British Library and of the Science and Technology Library connected to the Victoria and Albert Museum; in the Netherlands, my research centered on the collections of the Boerhaave Museum in Leiden and of the free library of Amsterdam. Visits to the Royal Institution in London, arranged by Dr. Frank Jones, and to the Kamerlingh Onnes Laboratory at Leiden, arranged by director Dr. Jos de Jongh and emeritus director Dr. Rudolf de Bruyn Ouboter, were especially helpful in gleaning details about the settings in which James Dewar and Heike Kamerlingh Onnes worked.

I owe debts of gratitude to Coleman Hough for basic research and insights, to editor Laura van Dam for her enthusiasm and her ability to ask the best questions, to Steve Fraser for getting me going on the book, and to my wife, Harriet Shelare, and sons, Noah and Daniel, for putting up with my obsession with an esoteric topic. Rudolf de Bruyn Ouboter, Russell Donnelly, and other physicists read portions of the manuscript and made helpful suggestions. The errors that may remain are, of course, mine alone.

Notes

The sections that follow report the major published and unpublished sources for the material in the book, with a few comments on them. The list is cumulative; that is, sources used in earlier chapters are also used in later ones, but I have omitted second references to them to make it easier to read. Though the listing does not constitute a complete bibliography, I hope it will provide ample fodder for readers who want to find out more about the people, events, history, and science of the cold.

Chapter 1

Any science-history research begins with the multivolume *Dictionary of Scientific Biography*, in whose pages the work of most (though not all) of the people discussed in this book are profiled. Gerrit Tierie's brief *Cornelius Drebbel (1572–1633)*, 1932, collects all the comments made by Drebbel's contemporaries and quotes liberally from his works. Thomas Tymme's *A Dialogue Philosophicall . . .*, 1612, provides the best description of Drebbel's fabled perpetual-motion machine. Two thoughtful articles are L. E. Harris's "Cornelius Drebbel: A Neglected Genius of Seventeenth Century Technology," *Newcomen Society*, 1958; and Rosalie L. Colie's "Cornelius Drebbel and Salomon de Caus: Two Jacobean Models for Salomon's House," *Huntington Library Quarterly*, 1954–1955. Fascinating references for the period are Lynn Thorndike's monumental *History of Magic and Experimental Science*, 1923; William Eamon's study of books of secrets, *Science and the Secrets of Nature*, 1994; and Elizabeth David's *Harvest of the Cold Months*, 1994. The last traces the work of della Porta and other alchemists and engineers mentioned in the chapter. Material about James I is provided in Robert Ashton's compilation *James I by His Contemporaries*, 1969, and in biographies, the most useful being Antonia Fraser's *King James VI of Scotland and I of England*, 1974. Westminster Abbey is interestingly traced in Edward Carpenter's *A House of Kings*, 1966.

Chapter 2

Barbara Shapiro's *Probability and Certainty in Seventeenth-Century England*, 1983, puts the Bacon-Boyle era, and its science, into perspective. *The Works of Francis Bacon*, in seven volumes, with notes by his disciples, was published between 1857 and 1859. Robert Boyle's *New Experiments and Observations Touching Cold*, 1665, is still a treat to read. Among the biographies of Bacon, Catherine Drinker Bowen's *Francis Bacon, The Temper of a Man*, 1963, slightly updated in 1993, is the most insightful, though it pays less attention to the scientific side than to the political. Steven Shapin and Simon Schaeffer's *Leviathan and the Air Pump*, 1985, analyzes the acerbic exchanges of Hobbes and Boyle; those authors' view of Boyle is countered in the best recent biography of Boyle, Mary-Rose Sargent's *The Diffident Naturalist*, 1995. The definitive study *The Royal Society: Concept and Creation* is by Margery Purver, with an introduction by Hugh Trevor-Roper, 1967.

Chapter 3

W. E. Knowles Middleton's books *A History of the Thermometer and Its Use in Meteorology*, 1966, and *The Experimenters, A Study of the Accademia del Cimento*, 1971, are exhaustive and thoughtful; a supplement is Maurice Daumas's *Scientific Instruments of the Seventeenth and Eighteenth Centuries*, 1972. Other references include Harold Acton's *The Last Medici*, 1932, and Christopher Hibbert's *Rise and Fall of the House of Medici*, 1974. Fahrenheit's 1729 letter to Boerhaave is put into context by the rest of the series, annotated by Pieter van der Star, in *Fahrenheit's Letters to Leibniz and Boerhaave*, 1983. Detective work on Fahrenheit's scale can be found in various articles in *Isis* and in *Nature*. The topic of "Antecedents of Thermodynamics in the Work of Guillaume Amontons" is analyzed by G. R. Talbot and A. J. Pacey in *Centaurus*, 1971, and by Robert Fox in *The Culture of Science in France, 1700–1900*, 1992, which also recounts the history of the Académie des Sciences. Robert Hooke's work is the subject of a lecture by E. N. da C. Andrade before the Royal Society, printed in its *Proceedings*, 1950. *Anders Celsius*, a biography by N. V. E. Nordenmark, was issued in 1936.

Chapter 4

Richard O. Cummings's *The American Ice Harvests*, 1949, and Oscar Edward Anderson, Jr.'s *Refrigeration in America*, 1963, are invaluable, as are Xavier de Planhol's *L'Eau de Neige*, 1995; Roger Thevenot's *A History of Refrigeration Throughout the World*, 1987; and W. R. Woolrich's *The Men Who Created Cold*, 1967. Early refrigeration machines are detailed in Edward W. Bryn's *The Progress of Invention*, 1900, and in Robert Maclay's chapter, "The Ice Industry," in Chauncey Depew's *One Hundred Years of American Commerce*, 1895. Henry G. Pearson's seminal paper "Frederic Tudor, Ice King," which quotes liberally from Tudor's diaries, is in the *Proceedings of the Massachusetts Historical Society*, 1933. Several articles about Gorrie and Twining are in the pages of *Ice and Refrigeration* and *The Florida Historical Quarterly*. Faraday's experiments are detailed in John Meung Thomas's *Michael Faraday and the Royal Institution*, 1991.

Chapters 5–6

Michel Serres sets the historical scene in "Paris 1800" in his *A History of Scientific Thought*, 1995. Robert Fox, in *The Caloric Theory of Gases*, 1971, traces the history of that wonderfully misleading concept. Sadi Carnot's *Réflexions sur la puissance motrice du feu*, 1824, is still in print. Hippolyte Carnot's memoir of his brother, along with Sadi's beautifully handwritten post-1824 notes, 1878, make fascinating reading. Good secondary sources are *Sadi Carnot, Physicien et les Carnots Dans L'Histoire* by A. Friedberg, 1978; *Carnot et la Machine a Vapeur* by Jean-Pierre Maury, 1986; and two articles by Robert Fox on Carnot, Clément, and work on steam engines, reprinted in his 1992 book *The Culture of Science in France, 1700–1900*. *Robert Mayer and the Conservation of Energy*, by Kenneth L. Caneva, 1993, tells more than anyone might want to know about the enigmatic doctor. *Scientific Papers of James Prescott Joule*, 1887, are now more available, thanks to a recent reprinting. *James Joule, A Biography*, by Donald S. L. Cardwell, 1989, and Cardwell's earlier *From Watt to Clausius*, 1972, are essential reading about the history of thermodynamics, as is Crosbie Smith and M. Norton Wise's *Energy and Empire*, 1989, the best biography of Lord Kelvin. Two other studies are Harold I. Sharlin's *Lord Kelvin, The Dynamic Victorian*, 1979, and David B. Wilson's *Kelvin and Stokes*, 1987. For those undaunted by mathematics, there is C. A. Truesdell III's *The Tragicomical History of Thermodynamics, 1822–1854*, 1980, and, for unrivaled clarity, various articles and book chapters on the same subject by Crosbie Smith. Daniel D. Pollock's article "Thermoelectricity" in the *Encyclopedia of Physical Science and Technology*, 1987, evaluates Thomson's contributions to the subject, and Joule-Thomson cooling is historically traced in Graham Walker's *Miniature Refrigerators for Cryogenic Sensors and Cold Electronics*, 1989. Bernice T. Eiduson's study *Scientists: Their Psychological World*, 1962, offers insightful observations.

Chapters 7–12

The clearest and most cogent book to deal with the entire subject of gas liquefaction and superconductivity is the second edition of Kurt Mendelssohn's *The Quest for Absolute Zero*, 1977. Other essential texts for this period are Per F. Dahl's *Superconductivity: Its Historical Roots and Development*, 1992; Gianfranco Vidali's *Superconductivity: The Next Revolution?* 1993; and Ralph G. Scurlock's *History and Origins of Cryogenics*, 1993. A supplement is Jean Matricou's *La Guerre du Froid*, 1994.

Cailletet's communications to the Académie des Sciences are salted through several volumes of its *Comptes Rendus* and its *Annales de Chimie*. Maurice P. Crosland's *Science Under Control: The French Academy of Sciences, 1795–1914*, 1992, is the definitive study of that institution during its most distinguished and influential period. Additional insight is provided by Crosland's *Studies in the Culture of Science in France and Britain Since the Enlightenment*, 1995. Gwendy Caroe, who spent her youth living in the Royal Institution, has written the best short book on it, *The Royal Institution, An Informal History*, 1985.

The Collected Papers of Sir James Dewar, in two volumes, 1927, was edited by his wife and several colleagues and includes such matter as newspaper reportage of his speeches where no other written material exists. Important supplements are articles by Agnes M. Clerke and by Henry E. Armstrong in the *Proceedings of the Royal Institution*, 1901 and 1909, which analyze the low-temperature research. Armstrong's longer memoir, *James Dewar*, 1924, is the most significant source of biographical information. Morris Travers's two books, *The Discovery of the Rare Gases*, 1936, and *A Life of Sir William Ramsay*, 1956, discuss the Ramsay-Dewar controversies. Tadeusz Estreicher's chapter "The Siamese Twins of Polish Science," referring to Olszewski and von Wróblewski, is in *Great Men and Women of Poland*, edited by Stephen P. Mizwa, 1942.

Carl Linde's autobiography *Aus Meinem Leber und Meinem Arbeit*, 1979, is explicated in Mikael Hard's *Machines Are Frozen Spirit*, 1994. Robert J. Soulen wrote about Dewar's flask in *Physics Today*, 1996. Some of the Dewar-Onnes correspondence is printed in an English edition of the selected papers of the *Communications from the Physical Laboratory at Leiden*, entitled *Through Measurement to Knowledge,* edited by Kostos Gavroglu and Yorgos Goudaroulis, 1991. More of it exists in manuscript at the Boerhaave Museum in Leiden. Anne C. Helden's pamphlet "The Coldest Spot on Earth," 1989, is a reliable guide to Onnes's procedures. Additional light is shed on Kamerlingh Onnes's contributions by Rudolf de Bruyn Ouboter in various recent articles, and in a chapter of the Gavroglu book. Leo Dana's memoir of his year with Onnes is in *Cryogenic Science and Technology,* edited by R. J. Donnelly and A. W. Francis, 1985. Another good source of biographical information is a memorial lecture by Ernest Cohen in *Journal of the Chemical Society,* 1927. The discovery of superfluidity is well covered in an article by Donnelly in *Physics Today,* July 1955.

The work and lives of Pyotr Kapitsa and Lev Landau are best traced, in English, by a translation of Anna Livanova's 1980 memoir *Landau, A Great Physicist and Teacher,* and by such books as *Kapitza: Life and Discoveries,* 1984, by F. B. Kedrov, and *Kapitza, Rutherford, and the Kremlin,* 1985, by Lawrence Badash.

Chapter 13

In *Frozen Foods: Biography of an Industry,* 1963, E. W. Williams tackles the subject with an insider's knowledge, as does a lecture by Clarence Francis of General Foods, "A History of Food and Its Preservation," 1937, which details the work of Clarence Birdseye. Gail Cooper's *Air-Conditioning America: Engineers and the Controlled Environment, 1900–1960,* 1998, deals properly with a neglected subject. The growth of the cryogenics industry and of the commercial use of liquefied gases is traced in David Wilson's *Super-cold,* 1979. In *Biophysics and Biochemistry at Low Temperatures,* 1986, the topic is examined by Felix Franks. The "Woodstock of Physics" era and high-critical-temperature superconductivity are the subjects of a handful of books, the best of them being Bruce Schecter's *The Path of No Resistance,* 1989, and Robert M. Hazen's *Breakthrough,* 1988. Articles about superconductivity in the decade after the 1987 breakthrough are found in issues of *Scientific American, Science, Physics Today,* and other journals, as well as in the science sections of the *New York Times.* The nexus of low-temperature physics, particle physics, and astronomy is detailed in *Near Zero: New Frontiers of Physics,* edited by J. D. Fairbank, et al., 1988, and especially in William Fairbanks's chapter, "Some Thoughts on Future Frontiers of Physics." Information on commercial work in superconductivity and liquefied gases comes from Web sites of the United States Department of Energy and the various industrial companies involved in these endeavors, and information on recent advances in various fields utilizing low-temperature research is taken from the Web sites of MIT and other universities and from newspaper reports. Contemporary work on cooling techniques for electronic equipment is gleaned from Win Aung's edited compilation for the National Science Foundation, *Cooling Techniques for Computers,* 1991; *Thermal Measurements in Electronics Cooling,* edited by Kaveh Azar, 1997; and the proceedings of the American Institute of Physics, Thirteenth International Conference on Thermoelectrics, 1995.

Index

1592

Galileo invents
the thermoscope,
an early form
of thermometer.

1620

Cornelis Drebbel
attempts to
air-condition
Westminster Abbey.

1657

Grand Duke
Ferdinand de Medici
develops
the first accurate
liquid-in-glass
thermometer.

1665

Robert Boyle
publishes *New
Experiments and
Observations Touch-
ing Cold,* which
dispels ancient myths
and reveals many
facts about the cold.

1703

Guillaume Amontons
mathematically
derives the
idea of an
"absolute zero."

1720

Daniel Fahrenheit
invents his
thermometric
scale.

1741

Anders Celsius
develops
the centigrade
scale.

1748

William Cullen
produces artificial
refrigeration
in the
laboratory.

1787

Martinus van Marum
liquefies ammonia
at a temperature
well below freezing.

1800

Thomas Moore
builds the
first portable
"refrigerator," lined
with rabbit fur.

1823

Michael Faraday
liquefies chlorine
and ammonia and
recognizes that
the chemical change
from gas to liquid
state radically
lowers temperature.

1824

Sadi Carnot
publishes a thesis
that founds the field
of thermodynamics,
the study of heat
and cold.

1834

Charles Saint-Ange
Thilorier concocts
"dry ice" and
reaches −110°C.

1834

Jean-Charles-Athanase
Peltier discovers
thermoelectricity, a
new way of
producing heat
and cold.

1848

William Thomson
(later knighted
Lord Kelvin)
develops his
absolute temperature,
or Kelvin, scale.

1850–1860

William Thomson
and Rudolf Clausius
separately articulate
the first and
second laws of
thermodynamics.